Rootedness

Rootedness

THE RAMIFICATIONS OF A METAPHOR

Christy Wampole

The University of Chicago Press CHICAGO & LONDON

CHRISTY WAMPOLE is assistant professor of French at Princeton University.

The University of Chicago Press, Chicago 60637
The University of Chicago Press, Ltd., London

Printed in the United States of America

25 24 23 22 21 20 19 18 17 16 1 2 3 4 5

ISBN-13: 978-0-226-31765-6 (cloth)
ISBN-13: 978-0-226-31779-3 (e-book)
DOI: 10.7208/chicago/9780226317793.001.0001

The University of Chicago Press gratefully acknowledges the generous support of the Department of French and Italian at Princeton University toward the publication of this book.

Library of Congress Cataloging-in-Publication Data
Names: Wampole, Christy, 1977– author.
Title: Rootedness : the ramifications of a metaphor / Christy Wampole.
Description: Chicago ; London : The University of Chicago Press, 2016. | Includes bibliograpical references and index.
Identifiers: LCCN 2015034628 | ISBN 9780226317656 (cloth : alk. paper) | ISBN 9780226317793 (e-book)
Subjects: LCSH: Roots (Botany) in literature. | Metaphor in literature. | French literature—20th century—History and criticism. | German literature—20th century—History and criticism. | Alienation (Philosophy) | Affiliation (Philosophy)
Classification: LCC PQ307.R63 W36 2016 | DDC 840.9/353—dc23 LC record available at http://lccn.loc.gov/2015034628

♾ This paper meets the requirements of ANSI/NISO Z39.48-1992 (Permanence of Paper).

To Iva, Alice, Carl, and Sue Ellen.
Thank you for raising me in a flower shop.

A psychologist conducting a lengthy study of different images of the root would explore the human soul in its entirety. A whole book could be written on this theme.

GASTON BACHELARD

CONTENTS

ACKNOWLEDGMENTS

I would like thank all of my colleagues in the Department of French and Italian at Princeton University for their support throughout this project. In particular, David Bellos, Nick Nesbitt, Efthymia Rentzou, Göran Blix, Gaetana Marrone-Puglia, Florent Masse, Tom Trezise, and François Rigolot have made great efforts to mentor me as a rookie professor. I would also like to thank André Benhaïm, Marie-Hélène Huet, Natasha Lee, Volker Schröder, Katie Chenoweth, Simone Marchesi, Pietro Frassica, Christine Sagnier, Fiorenza Weinapple, Murielle Perrier, Sara Teardo, Rachel Hart, Topher Davis, and Christophe Litwin, as well as all of the graduate students and lecturers and everyone else in the department for being fantastic colleagues. I'd like to extend a warm thanks as well to the administrative staff, including Ronnie Pardo, RuthAnne Lavis, Kathy Varra, Kathleen Allen, Crystal Arrington, Mike Rivera, and Sam Evans. They make coming to work a real joy. I'd also like to thank Sarah Kay for her guidance before my arrival at Princeton.

I have met so many wonderful, supportive friends and colleagues outside my department and at other institutions, and I want to thank them as well. Many attended the work-in-progress lecture where I presented this book in its embryonic stage; they provided extremely helpful input as it came into being. Others read my pages later or were simply there for me when I needed advice or a good conversation. I'd especially like to thank the following people for their support: Jill Dolan, Gerald Prince, Jack Woods, Maurie Samuels, Russ Leo, Tom Kavanagh, Emily Apter, Rava da Silveira, Emmanuel Bourbouhakis, Gabriella Nouzeilles, John Logan, Nigel Smith, Eduardo Cadava, Hannah Freed-Thall, Dora Zhang, Barbara Vinken, Howard Bloch, Anselm

Haverkamp, Tom Harrison, Florian Klinger, Morgane Cadieu, Alice Christensen, Mareike Stoll, Peter Brooks, Eileen Reeves, Nikolaus Wegmann, Arnaud Perret, Yue Zhuo, Moussa Sow, April Alliston, Carol Jacobs, Jeanette Patterson, Sarah Chihaya, Nefin Dinç, Jessica Devos, Scott Francis, Stefano Ercolino, Daniel Heller-Roazen, Jeff Dolven, Michael Hoyer, Barbara Nagel, Susan Stewart, Tom Connolly, Rubén Gallo, Henry Sussman, Dan Hoffman-Schwartz, Rüdiger Campe, Dan Edelstein, Dylan Montanari, Carol Rigolot, Mary Harper, and Devin Fore. Thanks also to Caley Horan; without her unwavering friendship, the ride would have been a bumpy one.

I wrote most of this book in the Irving S. Gilmore Music Library at Yale University during my sabbatical year there. I'd like to thank the librarians and the friendly security and café staff there, particularly Kenny, Mike, and Frank, who were witnesses to my long hours and who always had a kind word of encouragement after an exhausting day. Also, I'd like to give a special thanks to the Department of French and Italian and the Division of Literatures, Cultures, and Languages at Stanford University, the French departments at Yale and the University of Pennsylvania, and the Yale German department.

In the fall of 2014, I taught a seminar cross-listed in the French and German departments at Princeton entitled "Roots in Twentieth-Century France and Germany," which was based on all of the research I'd accumulated for this book. I want to give a very special thanks to the students who participated in this seminar, namely, Austin Gengos, Adeline Heck, Assimakis Kattis, Andrew Nelson, Liana Pshevorska, Alex Raiffe, and Marie Sanquer. The dynamic in the class was almost familial, and I am honored that the students participated so enthusiastically in the course. I don't recall any other class that inspired me quite the way this one did. I want to thank Natalie Berkman as well, who took the time during her year abroad in Paris to dig in the Bibliothèque Nationale for me.

Without the guidance of Alan Thomas at the University of Chicago Press, this book would never have reached the public. I appreciate his tireless efforts in walking me through the publication process and his careful attention to my pages. I also want to express my gratitude to Randy Petilos, who helped put the finishing touches on the manuscript.

This project would not exist without the inspiration of Robert Pogue Harrison, my adviser at Stanford, who has since become my friend. If I were asked to trace my intellectual lineage, it would always lead back to him. Anyone familiar with his work will be able to identify my book as a very Harrisonian project. I'd like to thank him for his warmth and care for the duration of my project and long before. He read many drafts of my chapters with careful attention and had priceless advice to offer. Our shared affinity for all things botanical resides in each page of this book.

There are several women who have mentored me and helped me to get where I've gotten in my personal and professional endeavors. These include my great-grandmothers, my grandmothers, and my mother, who have all been strong role models throughout my life; my high school French teacher Dede Hart; my undergraduate French professor Isabelle DeMarte; and Samia Kassab-Charfi of the Université de Tunis, whom I met at Stanford. I'd like to give an extra warm thanks to Marielle Macé, whose work has been very motivating for my own and whose kindness as an interlocutor is unparalleled.

I give a warm thanks to my family, who have supported me despite my incorrigible nomadism. And several dear friends—Steffen, Ulrich, Christophe, Monika, and Hannes—have kept life special during the process of writing this first book. I thank Karen and Marc in particular for existing; a world without them would be an impoverished one. There are two people who carry me on their shoulders, with love, with enthusiasm, and with the deepest affection. Beth, my best friend and the most generous person I know, has watched this book grow into what it has become. I thank her for her love and care and feel lucky every day that I have her in my life. And I thank Florian, the person who made things crystal clear, for the great and small gestures and the immense tenderness implicit in them all.

And, if it is permitted to thank a place, I thank my hometown of Kennedale, Texas. The pasture behind my grandma's house, the house where she was born and where she still lives, is the only place I can imagine as the ground of my symbolic roots.

INTRODUCTION

To be radical is to grasp the root of the matter. But, for man, the root is man himself.

KARL MARX, *Critique of Hegel's Philosophy of Right*

In the Mariposa Grove of giant sequoias in Yosemite National Park, a new scale of life is given to visitors. These awe-inducing trees disregard the frame of the photograph by residing mostly outside of it. Many of them have been given names—Grizzly Giant, Faithful Couple, Bachelor and Three Graces—and their furlike bark sustains the impression that you are walking through a prehistoric landscape of quiet titans that might move at any second. Some of them have occupied the same plot of ground for three thousand years. The people who swarm at their bases, necks swiveled upward, will die before the trees do. A figure of living constancy, the same sequoia that astonished a child might impress her again when she returns to it in her retirement years. To behold the sequoias is to understand why trees have been taken up as sacred symbols in so many faiths across the planet. They resist the invisible gravitational force and act as living representatives of transcendent verticality. Something underground holds this spectacle in place.

Yet in the everyday, trees and vegetal life are nearly invisible in their less spectacular forms. You've likely crossed paths with them today and failed to register their presence consciously. No longer singular living entities but a green and quiet backdrop, they provide the context in which we live. Their leaves work as a seasonal clock. Their canopies give shade. They feed us, clothe us, house us. They are everywhere and thus nowhere.

To mark the beginning of life, trees are crafted into cribs and cradles and birth certificates. In a simple lullaby, the beginning and end of a human life can be imagined through the tree:

Rock-a-bye baby, on the treetop,
When the wind blows, the cradle will rock,
When the bough breaks, the cradle will fall,
Down will come baby, cradle and all.

In the formative years of a life, trees make books, classroom desks, and children's blocks. We learn through them. Architects sculpt them into houses and towns, the polis where people are governed. Trees mark the end of human life when the car forgets to stay on its custom-constructed turf, as was the case for Albert Camus. They are the sites of lynchings, the fuel for the burning of witches and martyrs. They are used to build crucifixes and pine boxes. The shroud is a product of the tree. The death certificate and the last will and testament are the final, paper-thin scraps of an existence fashioned from plant life.

*

And what of the root? The tree is one thing, the root quite another. More literally invisible than the trunk and branches, the roots are a site of extreme figuration. It is through the root that underground life is imaginable and where attachment takes a botanical form. People often imagine themselves just as radically embedded as the plants that blanket their home region. The person is marked by his or her place. People are convinced of the solidity of their family tree and its firm rootedness in time and place. Transmission of culture moves like sap through subterranean ducts. A first set of questions: Why do humans so readily botanify themselves, insisting they are rooted—geographically and culturally—to a specific set of coordinates? Why are exile, diaspora, emigration, and expatriation described in the language of uprootedness? Why is destruction also known as eradication? Why are family histories organized in the shape of a tree, and where exactly are its roots?

Rootedness is a primary organizing trope that accommodates the need to feel connected to something outside the self; for reasons I will explain, this subterranean, botanical form seems the ideal metaphor to communicate that desire. Across cultures and through time, the root surfaces again and again as a figure for filiation, cultural connectedness, regional or national allegiance, and symbiosis with the environment. We tend to think of language as something that sprouted from an original seed and continues to burgeon. Certain kinds of knowledge are readily organized in an arborescent shape. Something in our brains is compatible with the tree's style of proliferation.

It is fair to say that a large part of what is usually called the West is suffering from an overwhelming sense of rootlessness. But what is it? Simply a synonym for alienation or disconnectedness? A metaphorical way to say that people can no longer count on institutions they believed in and cannot depend on security, community structures, or even language? In any place that has been touched by large-scale displacements of people owing to war, colonization, famine, disease, or political ostracism; or where people have lost contact with ancestral memory; or where there is a general sense of isolation resulting from either new social habits or technology's alienating effects; or where people feel estranged from the environment because of the human's self-extraction from it, people describe themselves as uprooted. The sensation reaches far beyond the West. Is our species, the exception to so many rules, destined to live its fate of exception all alone, uprooted from nature, history, and existential serenity?

My objective is to understand why we think about existence and dwelling in terms of rootedness. Philosophers, mystics, poets, and scientists have all, in their own ways and through their own discourses, participated in the metaphorization of person as plant. The narrative I construct here shows how the paganistic idea of human embeddedness in nature gets refashioned time and again, lyricized through literature and legitimated through politics, philosophy, and science. There is something clearly so compelling in the root as a metaphor that it has ceased to be recognized as such. Its literalization begs the question, "Was it ever figurative, or is the human actually plantlike in some verifiable way?" Among the central problems I address in this book are the overlap of nationalism and ecology, the political uses of genealogy and etymology, the literalization of metaphors, and the overwhelming sense of alienation brought about by globalization and technological modernity. In surprising ways that I will point out, the root proves to be the solution again and again for describing and addressing these problems.

When I told people that I was working on a project about roots, the first association for many of them was Alex Haley's epic novel *Roots* (1976) and the subsequent film adaption (1977). This reflex to think toward Haley's book is meaningful; his title and its subject matter—the cultural heritage transmitted from generation to generation in the context of the American slave trade—gives voice to the legacy of a population torn from its homeland and sent toward an irrevocable and tragic future, without recourse to roots as landmarks. The relationship with one's ancestors is necessarily a problematic one. Although we didn't know them, we feel that perhaps something more than their genetic material resides within us, something deeper and more culturally consequential. The family tree, even when it is mythologized by anecdotes and stories projected onto pictures of our forebears, seems to contain unseeable truths. Our roots are where we sometimes look for explanations about our patterns of thinking and acting. We see them, perhaps erroneously, as sites for self-understanding.

As a person who left home and who plans never to return there to live, I must admit that this project is personal. In the dichotomy between *les nomades* and *les sédentaires*, I fall clearly and comfortably on the side of the nomads. And given the impact globalization has had on how we think about space, travel, property, and cultural mixity, my nomadism is congruent with the general tendency toward centrifugal living. But, in moments of homesickness, I feel prompted to interrogate my tendency toward departure while many people I know have never felt compelled to leave home. What defines the relationship a person has with home? This book is an attempt at an answer.

*

This formidable question could be approached in various ways. As a scholar of twentieth-century European literature and the history of ideas, I have chosen to focus on the clear epicenters of the root metaphor's transformation during this period: France and Germany. In these two places more than anywhere else, cultural debates organized themselves around the problem of roots and radicality. The appeal for a return to or a refusal of roots surfaces constantly and in unprecedented ways, owing to the particular interplay, in those nations, of nationalism, regionalism, Catholicism, residual paganism, Graecophilia, conservatism, reformism, anti-Semitism, xenophobia, technophobia, and colonialism. Twentieth-century France and Germany, at odds in countless respects, find consensus in their reliance on the solidity of rootedness as a concept, all the while customizing it to suit very specific political aims. The geographic proximity of these two countries and the richness and interconnectedness of their philosophical traditions account in part for this continuity. For example, the Cartesian schism of mind-body and culture-nature has had lasting impact in French and German thought since its conception, and these binaries surface again and again in discourses that try to mitigate rootlessness. The root metaphor, which some have argued is an archetypal image, shows its full range of volatility as it moves from biologic, positivist stylings, through nationalist, racializing discourses, to its current rhizomatic, poststructuralist treatment. One of the most fascinating recent developments in critical theory is the reintroduction of the idea that thought itself is botanical in nature and that humans should look toward plants as a model for dwelling in the world, because they provide an alternate paradigm for the treatment of space, time, consumption, and death. If we take this claim as a starting point and move backward in time, it becomes clear that France and Germany are its points of origin.

The reader will notice a richer discussion of French material than German material; this is due to the plain fact that I am a specialist in twentieth- and twenty-first-century French and Francophone literature. However, the importance of Germany in any account of the conceptual problem of rootedness in Europe cannot be understated. For this reason, I must consider both places as the core sites of my claims.

This is not a project about the symbolic meanings of trees and plants or their importance as metaphors in European thought, a theme that has already received a considerable amount of attention.[1] Instead, the book focuses on roots and, more specifically, rootedness. What does rootedness mean? How does this kind of attachment differ from other kinds? How does the root differ from other metaphors for origin? What does the constant use of this metaphor say about the human's relationship to the botanical world?

The relationship between Judaism and the trope of rootlessness is one of the most well-researched aspects of the root metaphor.[2] In addition, the identity politics stemming from the fallout of European colonialism, often borrowing metaphors of uprooting, transplantation, and vegetal invasion, have been

1. See, for example, Carey, *Tree*; Richard Hyman, *Trees: Woodlands and Western Civilization* (London: Continuum, 2007); Dumas, *Traité*; Owain Jones and Paul Cloke, *Tree Cultures: The Place of Trees and Trees in their Place* (Oxford: Berg, 2002); Elisabeth Bourguinat and Jean-Pierre Ribaut, eds., *L'arbre et la forêt: Du symbolisme culturel à l'agonie programmée* (Paris: Mayer, 2000); Gayle Samuels, *Enduring Roots: Encounters with Trees, History, and the American Landscape* (New Brunswick, NJ: Rutgers University Press, 1999); Laura Rival, ed., *The Social Life of Trees: Anthropological Perspectives on Tree Symbolism* (Oxford: Berg, 1998); Karla Schultz and Kenneth Calhoon, eds., *The Idea of the Forest: German and American Perspectives on the Culture and Politics of Trees* (New York: Lang, 1996); Harrison, *Forests*; and Jung, "Philosophical Tree."

2. Among recent studies on this topic, I am thinking specifically of Sarah Hammerschlag's excellent book *The Figural Jew*. Particularly in the chapters "Roots, Rootlessness, and Fin de Siècle France" and "The Ethics of Uprootedness: Emmanuel Levinas's Postwar Project," she gives a meticulous history of the metaphor, making a distinction between the rootless Jew as a myth and as a trope, and proposing the notion of "the figural Jew, an archetype for a new kind of difference in particularity whose function is to suggest that there is a positive moral valence to resisting the discourse of belonging that dominates both the universalist and the particularist versions of political identity" (18). See also Joyce Block Lazarus, *Strangers and Sojourners: Jewish Identity in Contemporary Francophone Fiction* (New York: Lang, 1999).

thoroughly analyzed in recent years.[3] In fact, very rich analyses of both issues, which fall under the rubric of multiculturalism, make up a large part of contemporary scholarship in twentieth-century French and German studies. While these themes return again and again in this book, I do not attempt to synthesize or make any claims about identity politics or enter into debates centered on multiculturalism. I am certain that specialists in these fields have already said more than I could hope to. Instead, my aim is to show that these iterations of the root metaphor participate in a larger pattern of the botanification of the human, which has culminated in recent studies claiming that humans and plants are connected through a complex cellular consciousness that disregards the boundaries of the body.

An immediate challenge I encountered as I began this project was that the use of the root as a metaphor for origin, beginnings, or genealogical heritage is so widespread in the languages of focus. I found endless banal uses of it by well-known writers and thinkers throughout the twentieth century. To prevent this from becoming a theme project that would involve merely listing conspicuous appearances of the term, I have been very careful in my selection of examples. If the word *root* could simply be replaced by *origin*, I generally dismissed it. The metaphor in this petrified form is neither interesting nor telling for the arguments I attempt to make here. Instead, I have selected examples in which the root metaphor is taken up to make the case—through the rhetorical mobilization of *ethos*, *pathos*, or *logos*, often all three simultaneously—that something in the human is undeniably plantlike.

Another challenge was to distinguish in the examples I've chosen between what might be universal in the metaphorization of the root and what is specifically French or German. A preliminary comment I would make is that nationality makes the biggest difference in the philosophical, spiritual, and political deployments of the root metaphor. For example, Heidegger's use of it is very specific to his milieu, which combines the lingering paganism of early Germanic culture, the nationally inflected belief in a superior race, and

3. See, for example, Christopher Miller, *The French Atlantic Triangle: Literature and Culture of the Slave Trade* (Durham, NC: Duke University Press, 2008); Richard Watts, *Packaging Post/Coloniality: The Manufacture of Literary Identity in the Francophone World* (Lanham, MD: Lexington Books, 2005); David Ohana, Claude Sitbon, and David Mendelson, eds., *Lire Albert Memmi: Déracinement, exil, identité* (Paris: Editions Factuel, 2002); Anne Malena, *The Negotiated Self: The Dynamics of Identity in Francophone Caribbean Narrative* (New York: Lang, 1999); Apter, *Continental Drift*; and Christopher Miller, *Nationalists and Nomads: Essays on Francophone African Literature and Culture* (Chicago: University of Chicago Press, 1998).

right-wing land worship—what we could call "conservative conservation"—of the National Socialists. Conversely, in lyrical or ecologically minded iterations, the root metaphor takes on a more universal character. I've tried to treat the problem of the universal and the particular on a case-by-case basis.

As is the case for any project of such magnitude, I've had to leave out many fascinating texts and authors. To prevent the book from resembling the tangled, vegetal proliferation it describes, it was necessary to restrict its scope, and I hope the reader will see this as a virtue. Undoubtedly, each person who reads it will think of examples from her or his own field of interest—which only attests to the significance of its claims.

At the end of my research, I was able to make the following general observations about the root metaphor:

1. Interest in one's roots increases in proportion to the perceived level of danger that threatens those roots. People think of themselves most as rooted when something (the foreigner, new values and technologies, forced expulsion) jeopardizes this perceived embeddedness in a culture or a place.
2. The desire for rootedness is a conservative desire for comfort. Comfort here means the existential consolation of strength-giving sameness.
3. Root-seeking is often symptomatic of a general sense of ecological alienation. It is the evidence of a species-wide guilt about our self-extraction from the Earth as system.
4. In the Western imagination, transcendence is predicated on rootedness.
5. The fear of rootlessness is the fear of decontextualization. Because the tree is both a living entity and a place, it allows for the imagining of life as a situated phenomenon. It is both contextualizing and contextualized.
6. The fear of abstraction often guides a person's search for roots. This is why celebrants of rootedness fear the future, the intellectual, digitized life, and the stranger: They are figures of the abstract, that which separates, distances, withdraws, or removes (*ab-stracts*) what is solid and real.
7. Root logic is selective. Atavism is appealing only when the ancestor can be positively mythologized. The shameful ancestor is left in the ground.
8. Proof of rootedness is necessarily impalpable. For example, the concept of *terroir* in its mythologizing iteration sets up conditions under which land imparts some elusive difference that only the initiated can detect. Folkloric etymologists point to the primordial—and thus unprovable—origins of words as proof of some arcane connection between the world and the word.

9. Root-seekers are inherently retrophilic. This love of the past and the belief that it was invariably better is the product of a nostalgia for what was not necessarily experienced personally. Rootedness is a necessary metaphor in grand narratives of cultural decline.
10. Because the root processes death into life, this resurrectional power allows people to imagine that an afterlife is possible, if only as a molecular transformation.

*

The narrative I build in this book must begin with the subconscious underpinnings of rootedness. For this reason, the first three chapters focus on the deep structures of the root metaphor, how it works, and what it communicates about the human. I begin by showing its archetypal features, the paradigmatic ways its power as an archetype has been consolidated, and the manifold instances in which Western culture relies on the trope of rootedness. The first three chapters thus focus on the metaphoricity of the metaphor. In chapters 4 through 7, history enters the stage. While equally invested in the way the root metaphor functions, these chapters contextualize its use by the most influential thinkers in France and Germany in the twentieth century. Organized more or less chronologically, these chapters show how the root-based mysticism and nationalism of the early twentieth century adopts the language of the rhizome by the end of the century, while still maintaining many of the same mystical elements and the same call for groundedness. If early-twentieth-century philosophers expressed a desire for human rootedness in paganistic terms, this is no less true of recent thinkers who have called for plant democracy. This book narrates the arc of attempts over the century to tweak and customize the root metaphor—typically toward political ends—all the while maintaining a strange continuity, struggling to recontextualize the human and restore the world's lost enchantment.

Laying the groundwork for the rest of the book, chapter 1 outlines the ways in which the root has been imagined as an archetypal image. I argue that the root is an absolute metaphor, in Hans Blumenberg's terms, emphasized particularly by his use of it in the very definition of metaphorology. I suggest that the root works as a kind of supermetaphor that has subsumed the most compelling aspects of three other metaphors for origin: the foundation, the source, and the seed. The root's metaphorization follows a few basic patterns: root as home, root as genealogical origin, root as miniature person, root as the past, and root as a severable connection with any phenomenon, particularly the environment. I claim that, in its various guises, the twentieth-century search for origins often projects a logic of filiation onto nonbiologic phenomena. Origins and transformations are described in hereditary terms as the family tree is mapped onto literary movements, etymology, philosophy, and many other fields. I show the general characteristics of a rooted conception of self as they manifest themselves in autobiography, biography, and biographical sketches of fictional characters, what I call radical portraiture. Through the observations of Carl Jung and Gaston Bachelard, I claim that the root is not only a powerful figure that represents home, the past, death, memory, and the mother; it is a figure for the subconscious itself. Its subterranean aspects and monstrous form have allowed for the proliferation of folkloric renderings, figuring prominently in the revival of the mandrake legend in twentieth-century German film. New cinematographic possibilities allow for the age-old Ovidian transformation from human to plant and vice versa to be shown with startling realism. The omnipresence of the root metaphor and the consistency of its patterns of use confirm its universal appeal; in the chapters that follow, its specificity in the context of France and Germany is analyzed.

Poetry typically takes up the root for its purposes in lyrical treatments of home, longing, memory, and family. In chapter 2, I look at four poets—Francis Ponge, Paul Celan, Eugène Guillevic, and Denise Levertov—who have used the root to think through four distinct problems: Ponge uses it to engage critically with the question of immobility, an important characteristic that distinguishes the person from the plant; Celan uses it as a figure of the absolute negation of a generation and of a people; Guillevic sees in the root's industriousness the starting point for a reflection on vegetal will; and Levertov appropriates it as a figure for ecological alienation. France's botanical poet Ponge displays an ambivalence about botanical immobility, imagining plants as passive captives but also as beautifully adapted entities that have evolved in such a way that the world must come to them. Celan's "Radix, Matrix" depicts the Shoah's neutering effect, having cut off the generative possibilities of an entire race. The poem performs the same severing and separating gestures as the event, the matricial eradication. Guillevic, the lyrical celebrant of Breton regionalism, finds in the root's diligence a model for humankind and projects his Leftist sympathies onto the hardworking root, the embodiment of radical trying. Finally, Levertov depicts humans who try to reroot themselves in the environment as existentially awkward beings, bereft of grace and unfit for nature. I explain how the failure of technology to cure the existential awkwardness described by Levertov is enacted in the botanical photography of Karl Blossfeldt, who tried to show that all artistic and architectural production could be traced to

Urformen, or originary forms, in the plant world. By cutting roots out of the picture's frame, Blossfeldt's camera enacts the same technological severing of ties with nature that is the general condition of modernity and postmodernity. Through close readings of each poet's handling of the root metaphor, I find that what connects them is the human's insatiable desire for context. To live without context is to live in an existential vacuum.

In chapter 3, I argue that conceptions of spiritual and intellectual transcendence in European thought are predicated on rootedness. Beginning with Plato's *Timaeus*, I show how the human has been imagined throughout history as rooted in the divine. The plant's verticality is conceived as a model for moral rectitude. Paul Claudel, the father of neo-Catholicism in twentieth-century France, uses the image of the crucifix rooted in the Garden of Eden as the starting point for a series of cryptographic tableaux meant to illustrate the necessity of spiritual rootedness. His multiplication and mashing up of metaphors anticipates the maximalist, imagistic combinatorics of Deleuze and Guattari, who made their claims in *Anti-Oedipus* and *A Thousand Plateaus*, not through reason but through the accumulation of synonymous images. Aside from their fascination with the vegetal world, these three writers share a penchant for mysticism. In the following sections of chapter 3, I show the ways in which the inverted plant, which thrives even when uprooted and replanted upside down, has been used repeatedly as a figure for intellectually purposeful unorthodoxy (in the case of the Encyclopédistes), for intellectual risk (Paul Valéry), and for the overturning of European values (Michel Tournier). Conversion is exemplified through the plant's unique ability to thrive with its branches in the soil and its roots in the open air. This chapter catalogs the varieties of transcendence that are possible through the root metaphor.

In chapter 4, I turn to the language of roots, which was in the air in the time of Simone Weil and Martin Heidegger. In the early twentieth century, the *Querelle du peuplier*—a debate between Maurice Barrès and André Gide, which gave voice to anti-Semitic, anti-intellectual, and anti-cosmopolitan tensions in France—had illustrated the extent to which roots were taken very literally by the more conservative thinkers of the period. From this example of French nationalism in its most reactionary form to the paganistic language used by the National Socialists for celebrating original autochthony, the appeal for a restored rootedness was everywhere. However, the two opposing worldviews that unfolded during the *Querelle* became synthesized in the late twentieth century, namely through concepts such as Jacques Derrida's *destinerrance*, Edouard Glissant's *errance enracinée*, Jean-Claude Charles's *enracinerrance*, and Vilém Flusser's "taking up residence in homelessness," all of which argue in favor of becoming rooted in rootlessness, embracing nomadic instability as the new default form of living. Weil and Heidegger could not have anticipated such a synthesis. The only solution they could propose for unstable Europe was the concerted effort to regrow individual and collective roots. In her book *The Need for Roots*, Simone Weil's primary objective was, I claim, the reversal of a course toward abstraction, represented for her by the capitalist worship of money, the mechanization of the means of production, the alienated worker and the *paysan*, Judaism and what she saw as the most abstract branch of mathematics, algebra. I juxtapose her solution for Europe's crisis with Heidegger's notion of a new groundedness necessary in the age of burgeoning technology. His formulation of the problem closely resembles the French term *terroir*, which suggests that land imparts specific qualities to agricultural products. Heidegger believed this was true of people as well; they and everything they produce culturally were ideally marked by the soil. Land left its trace on (agri)culture. His introduction of a mystical element to *terroir* brought his arguments very close to those of Rudolf Steiner and the anthroposophists, who envisioned a new kind of integral personhood that would recontextualize humanity. I show that Weil's anxieties about abstraction were more prescient than Heidegger's in their anticipation of the new virtualization of our own time.

The first part of chapter 5 discusses the lineage between Descartes's Tree of Philosophy, whose roots are metaphysics; Heidegger's call to attend to the soil that feeds the roots of this philosophy tree; and Sartre's chestnut roots, which jolt the protagonist of *Nausea* into the stark recognition of Being. Roquentin's apparent rhizophobia is actually a manifestation of his ontophobia, his fear of existence. In a close analysis of the novel, I show how Sartre dissolves the Cartesian subject and forces his protagonist to reckon with the fact that he shares something fundamental with the abject root: they equally exist and cannot escape existence, even through death. Sartre emphasizes the absurdity of this Being as much as Beckett did in *Waiting for Godot*, in which a single, contextless, and absurd tree adorns the stage. Eugène Ionesco's very definition of the absurd emphasizes the severing of transcendental, metaphysical, and religious roots. Sartre did not choose the site of Roquentin's crisis at random; for him, the tree had an extremely charged significance, which he communicates in his letters, diaries, and other philosophical writings. In short, he suffered from dendrophobia, the fear of trees. More broadly, the fear of vegetal life's encroachment on human terrain runs throughout Sartre's pages and through much of French literature of the late nineteenth and early twentieth centuries, evincing a fear that all human achievement will eventually be canceled out by the plant world, which far predated it. The second half of the chapter is devoted to the legacy of phenomenologists like Sartre, Husserl, Merleau-Ponty, Heidegger, and Voegelin, who insisted that philosophy needed to reroot itself and find a new ground from which to depart. Every experience of the world is necessarily discerned through the senses and the mind of a perceiving subject. This realization led the phenomenologists to identify the perceiving subject as the starting point for all philosophical reflection. Thought is thus rooted in perception. The language of root and ground that characterizes phenomenological thought is transmitted to those poststructuralist thinkers most heavily influenced by phenomenology.

Language, whose basic morphemes are known as word roots, is imagined to grow and proliferate like a plant. In this framework, I explain in chapter 6 how the root works as a figure of unity, the common denominator of various divergent lexical pathways. Relying on a vocabulary of the underground, of purity, and of genealogy, etymologists take as their task the retrieval of an irretrievable origin. Etymology, the often approximative search for the earliest lexical roots, became extremely politicized in twentieth-century Europe, both in its more scientific and in its folkloric iterations. As a tool of political manipulation, etymologies were often deployed to lay claim to cultural primacy, manifest, for example, in the struggle between Germanist and Romanist etymologists, who argued whether words were coined "through" Germanic culture or Latinate culture. The *Wörter und Sachen* movement sought to bind words to material objects, while other linguists tried to argue that the earliest syllables were experientially determined. A Cratylist revival had many arguing that words were not random, while others like Saussure argued that the relationship between sign and signifier was, in fact, arbitrary. Jean Paulhan, also skeptical of Cratylism, dedicated an entire book to discrediting those who had bought into the childish search for a mystical, primordial language. In it, he disputes the claim that the ancestors' words were superior or more pure than their contemporary forms. Jacques Derrida challenged the idea that language is rooted and worked actively to undo the passion for origin, while Maurice Blanchot pointed out the problems with conceptualizing language as a phenomenon of filiation. Such a view, he argues, is incredibly dangerous in that it allows one to think of history in terms of a neat, linear narrative, making it provisionally manageable but leaving aside the most inconvenient, messy, and yet essential aspects of the past. I close chapter 6 with Jean Starobinski's proposal that we cut the imagined umbilical cord connecting ancestors and descendants. He suggests that we observe our ancestors and our contemporaries from other cultures with the same detachment, since they are aliens to us in time and space, respectively.

Through the example of etymology, I show that the more untraceable a root is, the more it acquires a robustness and flexibility that allows it to be harnessed politically. The gesture of exhumation implied in etymological research is in reality a form of occultation.

Many have argued that the twentieth century represents a shift from a root-centered epistemology to a rhizome-centered one. In the concluding chapter, I analyze the geopoetics of Deleuze and Guattari's writing on the rhizome and show how their antigenealogical string theory, based on the strandular instability of the rhizome, is taken up by thinkers such as Edouard Glissant and Antonio Benítez-Rojo, who sought a language for describing the Caribbean—and, more specifically, the Caribbean plantation—as an identity laboratory of cultural combinatorics. I unearth the mystical current that runs through the root metaphors of Claudel, Weil, and Heidegger, through the rhizome metaphor of Deleuze and Guattari, to recent calls for a new plant-human symbiosis, which, despite borrowing a poststructuralist vocabulary, retains many of the same paganistic arguments used by earlier thinkers. In the fields of philosophy, anthropology, and botany, recent attempts have been made to argue for a more democratic treatment of botanical life based on a natural symbiosis between plants and people. The three books of focus are Matthew Hall's *Plants as Persons: A Philosophical Botany* (2011), Eduardo Kohn's *How Forests Think: Toward an Anthropology beyond the Human* (2013), and Michael Marder's *Plant-Thinking: A Philosophy of Vegetal Life* (2013). I show how a controversial precedent, *The Secret Life of Plants* (1973) by Peter Tompkins and Christopher Bird, is reworked by Hall, Kohn, and Marder using the legitimating discourses of their fields. The poststructuralist inflections in these books tend toward what could be viewed as neopaganism. I close with the claim that these recent theories try to attenuate some of the alienating effects of hyperrationalization and loss of spirituality by underscoring the root's transformation of death into life. This empirically verifiable process proves for rationalists that there is indeed an afterlife, if only a physical one.

CHAPTER 1

Welcome to the Rhizosphere

The day after Hurricane Sandy hit the East Coast of the United States on October 29, 2012, the entire region was strewn with uprooted trees. A walk in a nature preserve near my town in New Jersey took me past fallen, decades-old trees and their jutting roots. When trees this massive fall, they take with them a thick circumference of land, leaving holes that fill quickly with water. The roots I saw were never deep; they were gnarled and weblike, proliferating horizontally in an erratic network. Roots generally reveal traces of themselves at the base of trees, but merely to signal the extent of their unseen-ness. We are able to scrutinize roots only when they are decontextualized, exposed by weather, age, disease, or human transplantation. Catching sight of a full root happens only in instances of violence, infirmity, or death. This could account in part for the metaphorical darkness of our associations with the root, compounded by the literal darkness of its subterranean dwelling space. Roots commune with the dead. Their composition is reliant on decomposition. Botanists have given the name rhizosphere to the area of soil around the roots of plants, their subterranean dwelling. We will enter the rhizosphere now together.

This first chapter provides a broad overview of the way the root has been imagined. There are a few basic patterns of its metaphorization: root as home, root as genealogical origin, root as miniature person, root as the past, and root as a severable connection to any phenomenon, particularly the environment. Carl Jung and Gaston Bachelard recognized its power as a subconscious image, but the root is clearly a figure for the subconscious itself. This metaphor's broad applicability to such a variety of circumstances reveals a meaningful pattern in the human psyche: Continuity is preferable to discontinuity. The root is an integrating metaphor, one that allows for connections to be made between past, present, and future, between remote geographic spaces, between neighbors, between the human and its ecosystem. To be rootless is to lack context. What becomes clear in the incessant reliance on the metaphor of rootedness is that the desire for temporal, spatial, epistemological, and ontological continuity is an elementary human need. In the following pages, the chronicle of a species at odds with itself unfolds. Figurative language is the outlet for a deep-seated apprehension about permanent estrangement from the context whence we came.

SOME THOUGHTS ON METAPHOR

> Philosophers have been called poets. And both philosophers and poets live by metaphors. But their motives for making metaphors are not the same.
>
> STEPHEN C. PEPPER, "Philosophy and Metaphor"

> By their metaphors ye shall know them!
>
> GOTTFRIED BENN, *Primal Vision*

Geoffrey Gorer has argued that while Americans tend to use mechanical metaphors to describe "institutions and processes," Europeans rely constantly on arborescent ones.[1] In all its reductive wildness, this sweeping assessment caught the eye of the philosopher Hans Blumenberg (1920–1996) in his research for *Paradigms for a Metaphorology* (*Paradigmen zu einer Metaphorologie*, 1960). Gorer's claim was of interest to Blumenberg because of "the methodological import of the attempt to trace stylistic differences of a *way of life* back to a layer of elementary ideas that always shows itself most clearly where the 'supply of images' has been tapped."[2] Even in stating the objectives of metaphorology, Blumenberg uses root language: "Metaphorology seeks to burrow down to the substructure of thought, the underground, the nutrient solution of systematic crystallizations" (5). While many have focused on the second half of this formula and its emphasis on chemical processes, the full sentence and the larger context of the quote make it clear that Blumenberg conceives of the method of metaphorology as imitative of the root's siphoning of minerals from the soil. He proposes the notion of absolute metaphors, which "have a history in a more radical sense than concepts" (5). Blumenberg's

1. Gorer, *American People*, 151–52.
2. Blumenberg, *Paradigms*, 63.

translator Robert Savage pursues Blumenberg's insistence on the root theme in his afterword to the book, writing, "Metaphorology radicalizes conceptual history, in the literal sense that it directs its attention to the roots of concept formation."[3] When Savage uses the word "literal" here, he is attempting to make a distinction between the dictionary definition of *radical* and its empty, contemporary meaning that leaves out its etymological connection to plant life. But "literal" here still does not refer to the literal embedded structure that feeds and stabilizes plant life. Nowhere is there a literal root in any of Blumenberg's thought, but the metaphor is so stealthy that many fail to recognize it as one. In his methodology for the study of absolute metaphors, he argues that, in many cases, metaphors are not simply rhetorical flourishes, replaceable by nonmetaphorical language, but that they allow thoughts to be expressed that are impossible to express in nonfigurative terms. These are absolute metaphors, of which his primary example is that of light as knowledge, a *topos* throughout the history of Western thought. The root is clearly an absolute metaphor, a fact emphasized all the more by Blumenberg's use of it in the very definition of the discipline of metaphorology.

The nineteenth-century British statesman Lord Palmerston argued that "half the wrong conclusions at which mankind arrives are reached by the abuse of metaphors, and by mistaking general resemblance or imaginary similarity for real identity."[4] I will show that a literalization of the metaphor of rootedness has had very real political consequences, particularly in France and Germany, the epicenters of root thinking in twentieth-century Europe. The evocation of rootedness most often signals the conservative desire for a return to the past, imagined as more stable than the present. It is also a deferral to ancestral wisdom. This root may serve as a metaphorical anchor to a geographic space, a figure for cultural transmission from one generation to the next, or even a symbol of the subconscious; its structure approximates a map of countless unexploited possibilities latent in each set of circumstances a person encounters.

To understand in part why the root is such an effective metaphor for origin, we may compare it to three other metaphors of beginning: foundation, source,

3. Savage in ibid., 139.

4. He continues: "Thus, people compare an ancient monarchy with an old building, an old tree, or an old man, and because the building, tree, or man must, from the nature of things crumble, or decay, or die, they imagine that the same thing holds good with a community, and that the same laws which govern inanimate matter, or vegetable or animal matter, govern also nations and states." Palmerston quoted in Bulwer, *Life of Temple*, 2:261–62.

and seed. The first, an architectural metaphor, portrays the beginning as a human-made structure upon which everything else is built. The ideal foundation is solid and lasting, since the house built on a shaky foundation is bound to crumble. This metaphor is often used in philosophy—for example, in the long history of debates between foundationalists and antifoundationalists—because language and ideas are easily metaphorized into architectural forms (building an idea, bridging two concepts, deconstruction, etc.). The source, an aquatic metaphor, has more spiritually inflected connotations and also refers to the truth-dispensing original text.[5] The wellspring is the uncorrupted starting point, and although water is not itself alive, it contains and sustains life (as in the fountain of youth) and connotes purity (ablution). The seed, a compressed kernel of futurity, represents the beginning point of a vital thing; it is a figure of fertility. Compared in the New Testament to the Kingdom of Heaven, the mustard seed seems small but grows to great proportions. The root is a powerful metaphor because it groups in a single figure the primary features of the others: like the foundation, it embodies the stabilizing precondition for upward development; like the source, it is approximates water in its fluvial form and its function as a channel for fluids; and like the seed, it is a botanical figure for potential and growth. Combining the stability of architecture, the fluidity of water, and the vitality of plant life, the root is a kind of supermetaphor that subsumes the others. This flexibility manifests itself in fascinating places, for example, in its application to the seemingly oppositional categories of *jus sanguinis* (the right of blood) and *jus solis* (the right of soil). Since its founding as a republic, France moved from a conception of nationality based solely on bloodlines (*jus sanguinis*) to one that included birthplace (*jus soli*).[6] In both cases, the fact of birth is the determining factor.[7] The root metaphor

5. In Michel Charles's *L'arbre et la source* (Seuil, 1985), he uses the tree as a metaphor for the rhetorical tradition and the source as a metaphor for the commentative tradition, which he believes dominates the twentieth century. He argues that there are tree-oriented periods and source-oriented periods throughout the history of Western civilization.

6. Patrick Weil synthesizes the complex history of French nationality, writing, "La nationalité française apparaîtra au bout du compte . . . réaliser . . . le programme que Napoléon Bonaparte lui avait fixé: embrasser très largement, sans discrimination—par le sol, par la filiation, le mariage ou la résidence—le plus grand nombre de Français possible." Weil, *Qu'est-ce qu'un Français?*, 14.

7. Robert Harrison has made the etymological connection between birth (*naissance*) and nation, pointing out that Abraham Lincoln's Gettysburg Address describes "the founding of the republic as a birthing process." Harrison, "America," 68.

accommodates both juridical statuses: its vascular function can be compared to the genealogical transmission of *jus sanguinis*, and its embeddedness in the soil works effectively to represent the relationship between people and their land, as expressed through *jus soli*. Thus the shift from one administrative definition of nationality to the other did not require the disposal of the root metaphor. As an absolute metaphor, the root allows for the efficient saying of what is otherwise extremely difficult or even impossible to communicate. This chapter attempts to show the contours of this metaphor, which, although broad, has its limits.

Not only is it a particularly powerful subconscious image; I argue that it is a figure for the subconscious itself. The subliminal mind is often depicted as a subterranean network that follows a bifurcating logic rarely touched by the light of day. The root's hidden and abject form resembles the unsayable aspects of the psyche. This living botanical thing reaches for what it needs, formulating its desires by yearning for them. It is an embodied motivation. The plant's radical desire for what has decayed in the soil stands as a figure for memory, the reaching into dark recesses for what used to be alive. The root represents a matricial impression, an irrecuperable home where one's character and body were still in their embryonic phases. If people think of themselves as rooted beings, it is due to an umbilical memory of an attachment to the earth, a memory that has been severed in more ways than one. Something specific about the root encourages its metaphorization as both the tenor and the vehicle, to use I. A. Richards's distinction.[8] For example, when roots are anchors that help the tree remain steady, they are the tenor; when people say that they are rooted in their homeland, the root is the vehicle. The smoothness of our metaphorization of this botanical form makes us often forget that it is a figure of speech; we imagine ourselves to be as rooted as the oak tree outside the window. Because of its metaphorical flexibility, the root is a favored figure of poets, who use it often in lyrical works about exile, homesickness, filiation, and the symbiosis between humans and the earth. It is alive but cannot protest our metaphorization of it. Its ugliness and subterranean dwelling make it a suitable candidate for metaphoric appropriations related to the abject, the taboo, the illicit, the underground, and, indeed, evil itself. Furthermore, with its vascular design, the root resembles the veins, nerves, and neurons in human

8. See Richards's chapter "Metaphor," in *The Philosophy of Rhetoric* (London: Oxford University Press, 1936).

and animal bodies and the rivers and tributaries that saturate the land.[9] As I'll show, the application of this metaphor goes far beyond the morphological correspondence between the root and things of a similar form.

GENERATION RADIX

The late nineteenth and early twentieth centuries witnessed an obsessive, often scientifically driven search for origins or the most elementary forms of life and culture. As just a small sampling, take Darwin's *Origin of the Species*, Nietzsche's *Genealogy of Morals*, Jung's work on archetypes, Heidegger's *The Origin of the Work of Art*, Russell's *History of Western Philosophy*, Wittgenstein's *Remarks on the Foundations of Mathematics*, Lévi-Strauss's structural explorations of the origins of myths, Stephen Pepper's research into the root metaphors or ground metaphors on which all worldviews are predicated, Foucault's analyses of foundational power structures, and Derrida's extensive work on the origin of writing. In the 1920s, a Belgian priest and physicist named Monseigneur Georges Lemaître developed what has become known as the Big Bang Theory, which he dubbed "la théorie de l'atome primitif" (the theory of the primeval atom). In short, the twentieth century was very much about a search for collective roots. In the twenty-first century, with the explosion of genealogical research made possible by large ancestral archives, such as the Family Search website hosted by the Church of Jesus Christ of Latter Day Saints, and by genetic testing services provided by biotech companies like 23andMe, which allow for the tracing of one's ancestry using DNA samples, it is clear that the root quest mirrors the general inward and narcissistic turn of contemporary Western civilization.[10] The earlier question "How did we get

9. The term *dendrite*, which means treelike, is used in physiology to describe the branches of nerve cells. The term is also used to describe any treelike branching, found in crystals and snowflakes, for example. The thickest part of nerves and rivers is called the "trunk." The Strahler Stream Order, a hydrological classification system for describing the branching and hierarchy of tributaries, uses arborescent language (root, leaf, etc.) as well. The white, cerebellar matter in the brain is called the *arbor vitae* or "tree of life" because of its arborescent shape.

10. Regarding this culturally inflated importance of genealogical roots in the information age, Steven Pinker concludes, "Outside a small family circle, the links of kinship are biologically trifling, vulnerable to manipulation, and inimical to modernity." Pinker, "Strangled by Roots," 35. Hannah Arendt explains the quest for roots this way: "It is precisely because people are aware of the fearful distance that separates us from our beginnings that so many embark upon a search for the roots, the 'deeper causes' of what happened. It is in the nature of roots and 'deeper

here? Where did we come from?" has shifted, perhaps temporarily, toward a different question: "How did I get here? Where did I come from?" This search allows for the creation of a community of invisibles with whom you never have to interact.

Why do people feel such a strong connection with their ancestors? Why such pride or shame in forebears with whom you share genes and perhaps a language or a set of traditions but whom you'll never meet? Why do adopted children sometimes feel the need to find their biologic parents? This is one of the great mysteries of our species. Humans feel reassured by ancestral continuity. When they study their ancestors, they imagine themselves to contain some kind of residual essence of the dead. If a person comes from a long line of warriors, he might believe he embodies a warrior spirit. The shameful ancestor has caused many people to change their surnames to symbolically cut ties with the disgraceful progenitor. The family tree, the schematic depiction of blood relations, has both practical and mythological functions. Its scientific iteration, the genogram, allows for the study of hereditary patterns useful in the prognostics of disease, the analysis of family dynamics, and other biologic or psychological patterns from generation to generation. But the family tree also exists as a mythologized branching of the dead into the living. Brigitte Boudon has traced the family-tree imagery from early popular legends of the tree imagined as a father or a mother through the biblical tree of Jesse and finally to its contemporary format.[11] When the family lineage is corrupted through scandal, the death or murder of a meaningful relative, or the erasure of genealogical traces through slavery or genocide, the mourning over this corruption is frequently expressed in terms of root loss. This is clear, for example, in Paul Celan's poem "Radix, Matrix," which is analyzed in chapter 2. By giving the genealogical structure an arborescent form, the family can be imagined as an integrated whole, like a constellation that takes shape only when the human eye projects a particular configuration on a cluster of stars. The tree gives a narrative structure to blood relations, crafting a story out of arbitrary coital encounters. A certain causality is often projected onto the tree, for example in Emile Zola's gargantuan tales of inherited vices. So the family tree can offer shade and protection, but it can

---

causes' that they are hidden by the appearances which they are supposed to have caused. They are not open to inspection and analysis but can be reached only by the uncertain way of interpretation and speculation." Arendt, *Responsibility*, 260.

11. Boudon writes, "L'arbre-ancêtre, peu à peu dépouillé de son origine mythique, est devenu l'arbre généalogique, associé à la naissance, à la généalogie, aux cycles antérieurs de vie des individus et des communautés." Boudon, *Symbolisme*, 62.

also be the gallows from which you hang yourself. From the etymologically connected words *genesis*, *genocide*, *genealogy*, *genital*, *genius*, *generation*, *gentle*, *genre*, *gender*, *gentry*, *genuine*, *gentile*, *gentrify*, *generous*, *gene*, and *genus*, the working lexicon of root-bound thinking begins to coalesce.

HOME IS WHERE THE ROOT IS

The Congolese writer Alain Mabanckou, in his keynote address at the Twentieth- and Twenty-first-century French and Francophone Studies Conference in Long Beach, California, in 2012, told how he had wondered why in Europe and the United States, the umbilical cord is discarded after a baby is born. In the Congo, it is kept and buried in the soil of home, which remains a permanent place of return. With this symbolic attachment of the child to her homeland, a transfer from the mother's body to the body of Mother Earth, initiation into life begins with a tethering. Africa, the imagined cradle of humanity, is by no means the only place where roots and home are synonymous.[12] Despite the absence of such a ritual in Europe and the United States, the language of rootedness to one's home is omnipresent and seems a nearly universal trope across cultures.

I would like to begin with a very typical illustration of the language used to express one's rootedness to home, a first example of what I'll call radical portraiture, a biographical or autobiographical depiction of a person that relies heavily on tropes of rootedness to the land. This excerpt from Guy de Maupassant's *Le horla* influenced Maurice Barrès's *roman à thèse Les déracinés* (1897), a vindication of regional embeddedness.[13] The protagonist of Maupassant's novella declares in the first paragraph: "I love the country here, and I love living here because this is where I have my roots, those profound and delicate roots that attach a man to the land where his ancestors were born and died, and that attach him to what one should think and what one should eat; to customs as well as foods; to local idioms and peasant intonations; to the smells of the earth, of the villages, of the air itself."[14] A rich rereading of *Le horla* is

12. Jean-Claude Charles writes derisively of this conception of Africans as living evidence of humanity's primordial beginnings: *"Le nègre serait-il donc l'homme primitif, la souche de toutes les générations? Ne devient-il invariable dans ses premières habitudes, que pour conserver le type de notre première origine?"* Charles, *Le corps noir*, 143.

13. For more on this influence, see Jean Canu, "Maupassant et *Les déracinés*," *Revue d'histoire littéraire de la France* 41, no. 3 (1934): 448–51.

14. Maupassant, *Horla*, 3.

possible through a focus on this rhizocentric *incipit*, which could occasion an entirely different analysis of the Brazilian demon that arrives on the ship to disturb the protagonist's rootedness. This passage is a prime example of radical portraiture, whose typical elements—in biographical or autobiographical form, as fiction or nonfiction—include the blurring of the line between country as polity and country as landscape; a lexicon of attachment; an appeal to ancestral heritage; an insistence on specificity; sensorial enumerations; intimations of both the physical and the psychological associations with the homeland; and a hinting at an ineffable something that distinguishes this place from all others, a difference that language cannot accommodate. These characteristics are amplified in narratives in which the home is missing. Homesickness—known in French as *le mal du pays* and in German as *Heimweh*—is the greatest occasion for a reverence toward roots. Because the root is a figure of vital will and yearning—as it pushes through the soil, reaching for what it needs—the longing for home makes sense as a root image. Consider this poem by Robert Morgan called "Homesickness," in which the eyes of potatoes begin to reach, as a form of remembering:

> Potatoes left in darkness keep
> their eyes closed tightly, but
> potatoes touched by hint of light
> will sprout translucent shoots from eyes
> that stretch and feel around and crawl
> toward the illuminating source,
> as though remembering a home
> and luminous spark of vitality
> in distant world and distant time,
> across genetic memory
> and evolution's tangled tree,
> to germ of the original bloom
> across the gulf of history.[15]

The separation between home and here, between the ancestors and ourselves, is often an unbridgeable gap, which, however, does not prevent the impulse to recover points of connection with what has become remote; this seems to be an inherently human trait. Regarding the burial of the dead, Robert Harrison argues that "humans bury not simply to achieve closure and effect a separation

15. Morgan, *Terroir*, 34.

from the dead but also and above all to humanize the ground on which they build their worlds and found their histories."[16] But this separation from the dead at the time of burial is provisional; the rest of our lives are spent trying to recuperate a lost connection with them and with the earth, from which we feel permanently alienated. The "germ of the original bloom" and the "genetic memory" in Morgan's poem express a longing for existential rootedness and the return home. However, as the narrative of this study unfolds, we'll see that *Heimweh*, or homesickness, is always at odds with *Fernweh*, or wanderlust. The centripetal or centrifugal tendency of individuals and collectives turns sentimentality into a political dilemma.

Home is never neutral. In some cases, it symbolizes the corrupt site of painful memories, bad habits, outdated values, or traumas that were never given language. In Maryse Condé's polyvocal novel *Crossing the Mangrove* (*Traversée de la mangrove*, 1989), one of the characters claims that problems resemble trees because their roots are invisible.[17] When one character tells another that he wants to write a book entitled *Traversée de la mangrove* (the same as the title of the book the reader has in his or her hands), she shrugs her shoulders and replies, "You don't cross a mangrove. You'd spike yourself on the roots of the mangrove trees. You'd be sucked down and suffocated by the brackish mud" (158). The root is that which trips, as the mangrove roots jut from the surface, making it very difficult to cross the swamp. Consciously or unconsciously, Condé was pointing toward the etymology of the word *scandal*, which can mean a stumbling block or a snare. The entire novel is full of scandal after scandal, collective and individual, each emerging from the past to haunt various central and minor characters. Condé uses the root of the mangrove tree, a species native to Guadeloupe, as a figure for the arch-scandal, that hidden collective or individual history that catches the foot or drags the body of its victim to the underworld, showing that the past has the power to catch up with the present and pull it down and backward. The problems of today are extensions of the scandals of the past, particularly the original colonial scandal. Where can Caribbeans who lived during colonization and after it locate their roots? When home is destroyed, left behind, or conquered, the urgency to feel

16. Harrison, *Dominion*, xi.

17. "Life's problems are like trees. We see the trunk, we see the branches and the leaves. But we can't see the roots, hidden deep down under the ground. And yet it's their shape and nature and how far they dig into the slimy humus to search for water that we need to know. Then perhaps we would understand." Condé, *Crossing*, 139.

rooted increases exponentially. A significant pattern: Rootedness matters most when it is threatened.

JUNG AND BACHELARD GO DEEP: THE ROOT AS SUBCONSCIOUS IMAGE

The root is both an organizing image used by the subconscious and perhaps the most compelling figure *for* the subconscious. With its moisture, silence, darkness, minerals, and subterranean creatures, the rhizosphere is the home of the root and a kind of follicle or life-sustaining pouch. This subterranean home places the root in the archetypal category of the underworld. As a bridge between the Chthonic (the underworld) and the Uranic (the celestial), the tree, whose roots bury themselves in the earth and whose branches reach toward the heavens, has a privileged place in the human imagination as a sort of communicating vessel between the terrestrial world and the godly firmament. The imagined symmetry of the tree above and below ground transforms it into a kind of verticalized Rorschach test; our discernment of particular patterns in this symmetry and our interpretation of it says much more about us than it does about the tree. Roots are generally seen as the tree's stabilizing mechanism, the way it holds itself in place. But roots also prevent erosion by clutching the soil and holding it together. Metaphorically speaking, our roots keep us from losing ground.

If the subconscious is below, it shares this subterranean world with the root. Jung saw trees as "direct embodiments of the incomprehensible meaning of life"—which explains their importance in many of his most significant writings on archetypal imagery.[18] When he prepared a Festschrift for the seventieth birthday of the botanist Gustav Senn at the University of Basel, the theme he chose was the philosophical tree. In an expanded version of it, Jung shows the way the tree has been represented symbolically throughout history and how the inner experiences of his patients align perfectly with this cosmic pattern. He is careful to note that "none of the patients had any previous knowledge of alchemy or of shamanism,"[19] to underscore his argument that the human psyche has a universal archetypal structure into which all people are plugged, whether they have been schooled in the occult or not. The tree—roots included—is one of the fundamental images the human uses to make

18. Jung, *Earth*, 28.
19. Jung, "Philosophical Tree," 253.

sense of the world. Something about its elementarity and aliveness has encouraged philosophers, poets, and mystics throughout the centuries to use it as a screen onto which they can project any number of conceptual experiments.[20] The "wonder-working plant"[21] takes form throughout history as the Edenic Tree of Knowledge and Tree of Life, the Burning Bush in the Old Testament and the Torah, the larger-than-life Yggdrasil in Norse cosmology, the Christmas tree, the *Donareiche* of the Germanic pagan tradition, the sacred tree of Uppsala in Sweden, the Sephirotic Tree or the *Etz haChayim* (Tree of Life) found in the Kabbalah, the Ashvattha or Sacred Fig in Hindu mythology, and mystical trees in countless other traditions across the globe. In his 1976 book of poetry titled *Arbres* (Trees), Jacques Prévert melds magic and the tree in the clever portmanteau "Abracadabrarbre."[22] A tree is used in some cultures to represent the *axis mundi*, or the universal center where heaven and earth meet. Jung identifies numerous patterns that crop up in relation to the tree as archetypal image, noting that "the commonest associations to [the tree's] meaning are growth, life, unfolding of form in a physical and spiritual sense, development, growth from below upwards and from above downwards, the maternal aspect (protection, shade, shelter, nourishing fruits, source of life, solidity, permanence, firm-rootedness, but also being 'rooted to the spot'), old age, personality, and finally death and rebirth."[23] The root, that hidden twin tree underground, possesses most of these qualities in its own right, except the power to bear fruit or provide shade and protection. Jung cites the French poet Noël Pierre's poem "Soleil noir" as an "authentic experience of the unconscious" (270) that depicts the "universal midpoint of mankind" and "the heart as the centre of man's vitality in conjunction with the tree" (271):

> Quelle quiétude, au Nœud des Choses!
> Sous l'Arbre de ma vie, le Dernier Fleuve
> Entoure une Île où s'érige

20. From the twentieth century alone, we see Saussure's tree (*arbor*) that illustrates the difference between signifier and signified; Buber's tree as a starting point for exercises in empathy; Husserl's tree *simpliciter* that can be destroyed, though its concept cannot; and Sartre's chestnut tree that shows the revolting nature of Being.

21. Jung, "Philosophical Tree," 253.

22. Prévert, *Arbres*, 25.

23. Jung, "Philosophical Tree," 272. Regarding "personality," Jung writes, "There is a very ancient, indeed primitive idea that the tree actually represents the life of a man; for instance, a tree is planted at the birth of a child, and its and the child's fates are identical" (272–73n1).

Dans les brumes un cube de roche grise,
Une Forteresse, la Capitale des Mondes.[24]

(What quietness, at the hub of things!
Beneath the tree of my life, the last river
Surrounds an island where there rises
In the mists a cube of grey rock,
A Fortress, the Capital of the Worlds.)[25]

Jung's interest in this passage is due mainly to its power of synthesis. In one short text, Pierre manages to fuse many of the tropes that surround the root: its role as a kind of self-emanating center, its fluvial form and intimate connection with water, and its solitude and separation from the above-ground world. The root, as a hub or knot of everything, is the tangled place where unseen work happens. Jung claims that the "cube of grey rock" refers to the philosophers' stone, the alchemists' Holy Grail and path to immortality or at least rejuvenation. In his reading, the tree hides the philosophers' stone in its roots. The tree—and, more specifically, its secret repository of knowledge and life called the root—contains within it everything the human might need.

European philosophical and critical thought from the twentieth century abounds with botanical metaphors: Sartre's nausea-inducing root, Bataille's language of flowers, Paulhan's flowers of Tarbes, Derrida's *dissémination*, Deleuze and Guattari's rhizome, Cixous's pollen-gagged mouths, Irigaray's efflorescence, and Buber's tree of contemplation are just a few representative examples. Jean-Pierre Cléro has shown how the tree has been used extensively throughout the history of Western philosophy as a metaphor for reason.[26] Robert Dumas, in his book *Treaty of the Tree: An Essay on a Western Philosophy* (*Le traité de l'arbre: Essai d'une philosophie occidentale*, 2002), argues that the tree is a central, organizing figure for Western philosophy as a whole.[27] His many compelling examples include Saussure's use of the Latin *arbor* to illustrate

24. Pierre, *Soleil noir*, 41.

25. Jung, "Philosophical Tree," 271.

26. See Jean-Pierre Cléro, "L'arbre comme métaphore de la raison," in *Le végétale*, ed. Jean-Pierre Cléro and Alain Niderst (Rouen: Publications de l'Université de Rouen, 1999), 113–40.

27. "Nous avons voulu redonner à l'arbre la place qui lui revient dans la culture occidentale: la place centrale à partir de laquelle tout s'organise et prend sens." Dumas, *Traité*, 235. He writes that "la connaissance de l'arbre est la racine de l'arbre de la connaissance" (11).

the differences between signifier, signified, and referent; the persistent use of the tree graph to organize knowledge; and the speculative botany practiced by many of the most famous philosophers in the Western canon and its influence on their thinking in other disciplines. Elaine P. Miller, in her book *The Vegetative Soul: From Philosophy of Nature to Subjectivity in the Feminine*, shows the extent of German philosophy's reliance on plants as figures for thought and analyzes the implications of the tendency to gender the botanical as feminine and to "botanize" the female, with special focus on this theme in the writing of Kant, Goethe, Hölderlin, Hegel, Nietzsche, and others.[28] Michael Marder has published a catalog of various philosophers' uses of plants in his book *The Philosopher's Plant: An Intellectual Herbarium*, which dedicates particularly interesting chapters to Aristotle's wheat, Augustine's pears, Avicenna's celery, and Leibniz's blades of grass. These are just a few cases of the recent interest in the relationship between the realms of philosophy and botany, but this small sampling indicates how these fields are intertwined in often unexpected and fruitful ways throughout the twentieth century. More specifically, the root, not just the plant, has a lengthy and conspicuous history in European philosophical thought. French philosophy embraced the root as a noetic metaphor—a metaphor for consciousness, the mind, and ways of knowing—at least as early as the seventeenth century, when Descartes wrote in his *Principles of Philosophy* (1647) that metaphysics was the root of the Tree of Philosophy. In the twentieth century, Heidegger challenged Descartes's taxonomy, offering up his own metaphorical question: Metaphysics may be the root of philosophy, but what is the soil that nourishes it? Sartre also must have had Descartes's metaphysical root in mind when, in his *Nausea* (1938), the protagonist Roquentin's existential nausea is triggered most acutely by the visible root of a chestnut tree in the Jardin public. Sartre's novelization of the crisis of Western metaphysics participates in a general pattern in twentieth-century Europe of the dissolution of meaning, particularly after the world wars. Samuel Beckett toys with this idea as well in his *Waiting for Godot* (1952), in which a single tree adorns the stage, uprooted and contextless, a possible figure for the failure of Cartesian metaphysics.

28. Her analysis seeks "to investigate a facet of nineteenth-century *Naturphilosophie* and its antecedents that has been too little studied as a literary trope, and to suggest that the figure of vegetable genius transformed into the 'vegetative soul' might provide a fruitful alternative to the recently much-denigrated figure of the modern subject." Miller, *Vegetative Soul*, 200.

Jung was sensitive to the root metaphor and even used it in the title of one of his important texts on archetypes, *From the Roots of Consciousness: Studies on the Archetype* (*Von den Wurzeln des Bewusstseins: Studien über die Archetypus*).[29] Another term for the archetype is the *Urbild*, or originary picture, whose prefix *Ur-* will be encountered over and over again throughout my analysis: *Urszene* (primal scene), *Ursprache* (original language), *Urformen* (original forms), *Urpflanze* (original plant), *Urheimat* (original homeland), and, in its most basic form, *Ursprung*, or origin. These *Ur*-phenomena are frequently depicted through rhizocentric metaphors. The problem with the idea of an *Ur*-phenomenon is that it goes back so far into prehistory—like Goethe's imagined primeval plant from which all others derived, or like the first human utterances that were the obsession of many etymologists—that one can never be certain of its nature. The primeval instance becomes a blank root, a *radix rasa*, onto which almost any imagined origin can be projected. This projection, as I will show, is often politically motivated.

It is essential to notice that the archetypal image of the root (or any archetypal image, for that matter) poorly reflects the wide diversity of radical forms in extrapsychical life. The reductive archetypal image of the root represents a fraction of its possible structures in the natural world, for example, in its aerating, buttressing, propping, tuberous, or stilt forms or its aquatic or aerial types. Furthermore, the tendency to describe the root—as a figure for the past—as somehow older than the rest of the tree is also misguided. Plants constantly generate new cells, but for whatever reason, the root is imagined to predate the above-ground parts of the plant. Perhaps when a plant is seen sprouting from the ground, one imagines that a significant amount of unseen groundwork had to happen beforehand to facilitate this growth. Rather than imagining that the plant has a center point situated at the soil line from which everything grows centrifugally—as Noël Pierre did through his expression "le Noeud des Choses" (the "Knot of Things"), rendered by the translator as the "Hub of Things"—one imagines that the root builds itself below and then pushes a green finger up through the surface. In the Western imagination, the root is a bifurcating, underground, and messy reflection of the branches above and it is older than these branches. Archetypal images are always essentializing schemata, as is apparent in the discrepancy between the real and the imagined root.

29. Dominique Fernandez, member of the Académie française, also titled his book on psychobiography *L'arbre jusqu'aux racines: Psychanalyse et création* (1972).

Gaston Bachelard's 1948 *Earth and Reveries of Repose: An Essay on Images of Interiority* (*La terre et les rêveries du repos: Essai sur les images de l'intimité*) is the most concise catalog of the root's metaphorical and subconscious meanings. The word *intimité* in his title has several meanings, including "intimacy," "privacy," "the inner self," and even "comfort," and for this reason, "interiority" is perhaps the most inclusive translation that invites all of them. The book explores images that represent "rest, refuge, or rootedness"[30] and the psyche that folds in on itself.[31] Bachelard identifies certain images—such as the home, the cave, the womb, and the root—whose dynamic energy always tends toward rest.[32] Emily Apter argues, "It is literature in Bachelard that mediates and tames this archaic drive, sublimating ecodesire into a phenomenology of nostalgia for *la maison natale*."[33] Through poetry and fiction, these latent, primal energies can be vented. Subterranean life, imagined as the final resting place and a return to the womb of Mother Earth, is subconsciously strived for by all people. Both humans and plants rest in beds.

In his choice to elaborate on the root image, Bachelard shows his sympathies toward phenomenology; as I show later, this philosophy relies heavily on metaphors of rootedness and groundedness, insisting that the subject must recognize her or his embeddedness in experience. Borrowing Hans Carossa's notion of the human's piercing view, that penetrative gaze that looks beyond what is surface to seek the underground aspects of phenomena, Bachelard argues that the will to "violate the secret of hidden things" is part and parcel of the human's "aggressive, etymologically inspective curiosity."[34] He doesn't mean introspective, in the sense of looking into the self, but inspective: looking inside of things outside the self. To the human eye, objects have depths that ask to be explored. We thus feel sympathy with the root, whose vocation it is to probe the depths; it is, like us, an *être inspecteur*. The subconscious is conceived as below, sometimes architectonically (the brain as a building whose deep architecture, basements, cellars, hidden rooms, or *oubliettes* contain undiscovered truths, skeletons not in the ground but in the closet), sometimes

30. Bachelard, *Terre*, 5.

31. Bachelard uses the terms "psychisme involutif," "l'enroulement," and "involution" (5) to convey this inwardness.

32. On the separation between the human-made dwelling and the pre-architectural earthly dwelling, Levinas writes, "The home does not implant the separated being in a ground to leave it in vegetable communication with the elements. It is set back from the anonymity of the earth, the air, the light, the forest, the road, the sea, the river." Levinas, *Totality*, 156.

33. Apter, *Continental Drift*, 36.

34. Bachelard, *Terre*, 8.

geologically (an archaeological dig, as in Freud's famous opening lines about Rome in *Civilization and Its Discontents*).[35] Bachelard even uses both the root metaphor and the archaeology metaphor when defining Jungian archetypes.[36] He argues that the psychologist could get a very full picture of the soul based simply on the wide range of ways the root has been rendered by the imagination. For him, the tree is an *objet intégrant* (unifying object), that is, an object that allows for the integration of images (299). He describes the imagination itself as arborescent. His chapter is an accumulation of ways the root has been imagined by writers and thinkers through time and across cultures, with a strong focus on Europe. One of Bachelard's important contributions is his insistence on the problem of attachment. He links the remembered home of childhood to the crypt, the root, and other figures of depth, all of which have the power to tether.[37] The umbilical cord is never truly severed. We will all return whence we came: the ground. The root stands as a figure for limits. It represents the ways in which history reduces our range of experience and does not let go. Imagine a balloon tied to a guardrail: it has a range of movement and is made more dynamic in the breeze, but there is a certain perimeter beyond which the balloon cannot move unless its string is severed. In many depictions of the root, the bond with home or past is the absolute constraint. More common still is an ambivalent portrayal, alternating between sympathy and antipathy toward whatever is seen to be one's radical beginning, perceived at times as a stabilizing force and at other times as a handicap.

A final and significant subconscious association with the root is the problem of radical violence, in the form of root rot, eradication, or severing of the

35. Jung uses a similar metaphor to describe the psyche's structure: "It is as though we had to describe and explain a building whose upper story was erected in the nineteenth century, the ground floor dates back to the sixteenth century, and careful examination of the masonry reveals that it was reconstructed from a tower built in the eleventh century. In the cellar we come upon Roman foundations, and under the cellar a choked-up cave with neolithic tools in the upper layers and remnants of fauna from the same period in the lower layers. That would be the picture of our psychic structure." Jung, *Earth*, 68.

36. "Pour ce psychanalyste [C. G. Jung], l'archétype est une image qui a sa racine dans le plus lointain inconscient, une image qui vient d'une vie qui n'est pas notre vie personnelle et qu'on ne peut étudier qu'en se référent à une archéologie psychologique." Bachelard, *Terre*, 263–64.

37. "La maison du souvenir, la maison *natale* est construite sur la crypte de la maison onirique. Dans la crypte et la racine, l'attachement, la profondeur, la plongée des rêves" Ibid., 98.

root—all symbolically charged expressions that involve the disintegration of the tree's integrity. But cutting off a tree from its root does not necessarily kill it. Borrowing an image from the book of Job,[38] Toussaint Louverture used the root metaphor in 1802 as a rallying cry for the success of the Haitian Revolution, when he declared: "By overthrowing me, they have only brought down the trunk of the tree of freedom in Saint-Domingue; it will regrow because its roots are deep and many."[39] Jean-Bertrand Aristide, Haiti's first democratically elected president, later appropriated this declaration on the occasion of his own removal, in the first speech after his exile in 2004.[40] The root has the tenacity to survive its separation from the rest of the tree. When used metaphorically, this indicates a belief that the ancestors are potentially more resilient than their descendants. Such an atavistic logic suggests that even when traditions appear to have faltered, their hidden root thrives and is capable of flourishing unexpectedly. This atavism looks toward history, hoping that latent but surviving traces of a better past are still intact. The root metaphor, as we will see, is constantly evoked in situations of cultural rupture, ecological alienation, and technophobic angst. It gives voice to apprehensions about the human species as exception and, even within the species, about the inability to maintain continuity between generations or neighbors. In the coming chapters, I show the many forms radical rupture can take and the ways people respond with hope or hopelessness to this rupture using the language of roots.

38. The idea of a tree cut at its roots but still able to live is found in Job 14, verses 7–10, New International Version: "At least there is hope for a tree: If it is cut down, it will sprout again, and its new shoots will not fail. Its roots may grow old in the ground and its stump die in the soil, yet at the scent of water it will bud and put forth shoots like a plant. But a man dies and is laid low; he breathes his last breath and is no more."

39. "En me renversant, on n'a abattu à Saint-Domingue que le tronc de l'arbre de la liberté des noirs, mais il repoussera car ses racines sont profondes et nombreuses." Louverture quoted in Frères de l'Instruction Chrétienne, *Histoire*, 100. All translations are mine, unless otherwise noted.

40. "Dans l'ombre de Toussaint Louverture, le génie de la race, je déclare qu'en me renversant on n'a abattu que le tronc de l'arbre de la paix; il repoussera par les racines parce qu'elles sont louverturiennes." Aristide quoted in Munro, "L'exil," 142.

RADICAL EVIL: OF MANDRAKES AND WURZELMÄNNCHEN

Evil is often imagined as a rooted phenomenon.[41] Bachelard closes his reflections with a brief mention of the deadly mandrake (man + dragon, etymologically), the homunculus root he takes to represent the phallus.[42] The plant is

41. Examples include the commonplace that money is the root of all evil and Kant's notion of "radical evil" from his 1793 *Religion within the Limits of Reason Alone* (*Die Religion innerhalb der Grenzen der bloßen Vernunft*). Karl Jaspers writes, "There is a blight on the roots of man. A glance at history shows man struggling upward only to relapse; most of his high points were paid for in the cruelty and misery that made them possible. But the blight on the roots lies in every individual. We experience it in ourselves, as the 'radical evil' that overpowers us if we do not fight it constantly and freely." Jaspers, *Future,* 234. Hannah Arendt argued that evil is not radical, that it has no depth; but for this reason its nature as a surface phenomenon makes it capable of spreading, since roots cannot grow infinitely. She writes,

> It is of course true that evil was commonplace in Nazi Germany and that "there were many Eichmanns," as the title of a German book about Eichmann reads. But I did not mean this. I meant that evil is not *radical,* going to the roots (*radix*), that it has no depth, and that for this very reason it is so terribly difficult to think about, since thinking, by definition, wants to reach the roots. Evil is a surface phenomenon, and instead of being radical, it is merely extreme. We resist evil by not being swept away by the surface of things, by stopping ourselves and beginning to think—that is, by reaching another dimension than the horizon of everyday life. In other words, the more superficial someone is, the more likely he will be to yield to evil.
>
> *Jewish Writings,* 479

and

> The greatest evildoers are those who don't remember because they have never given thought to the matter, and, without remembrance, nothing can hold them back. For human beings, thinking of past matters means moving in the dimension of depth, striking roots and thus stabilizing themselves, so as not to be swept away by whatever may occur—the Zeitgeist or History or simple temptation. The greatest evil is not radical, it has no roots, and because it has no roots it has no limitations, it can go to unthinkable extremes and sweep over the whole world.
>
> *Responsibility,* 95

42. Jean-Luc Nancy has tried to offer a botanical alternative to the root's symbolic phallic intrusion with the vaginal metaphor of *béance* (gap or wide-openness) or *éclosion* (the opening of a flower or the spreading open of a larva) in his book *Dis-Enclosure,* the attempt at a preliminary deconstruction of Christianity:

> The eclosure (*l'éclosion*) of the world must be thought in its radicalness: no longer an eclosure against the background of a given world, or even against that of a given creator, but the eclosure of eclosure itself and the spacing of space itself. (In a sense, then, the word *radi-*

known for its potential to increase fertility but also to kill humans and dogs. The long cultural history of the mandrake includes its presence in the Rachel and Leah story in Genesis, its countless mentions in European texts on witchcraft and magic—particularly in the Middle Ages—for its healing and harmful properties, Shakespeare's repeated use of it, and its brief appearance in Beckett's *Waiting for Godot*.[43] This anthropomorphic root, a small vegetal member, can kill when it is uprooted, but it can also be a source of fertility. In some iterations, wrenching the little man from the earth is akin to castrating him. Albert-Marie Schmidt, in his book *La mandragore*, notes that humans have a tendency to equate themselves with plants.[44] This tendency makes the opposite projection—equating a plant with a human—extremely easy, especially in the case of the mandrake, whose form often resembles a small human with arms, legs, and a torso. The mythology of a miniature root person has had a particular hold on the German imagination. Many twentieth-century German-language children's books feature the Wurzelmännchen (the little root man).[45]

---

calness is inappropriate: it is not a question of roots, but of wide-openness [*béance*]). . . . Locations (*les lieux*) are delocalized and put to flight by a spacing that precedes them and only later will give rise (*donnera lieu*) to new places (*lieux nouveaux*). Neither places, nor heavens, nor gods: for the moment it is a general dis-enclosing, more so than a burgeoning. Dis-enclosure: dismantling and disassembling of enclosed bowers, enclosures, fences. Deconstruction of property—that of man and that of world."

Nancy, *Dis-Enclosure*, 160–61

43. Estragon: What about hanging ourselves?
Vladimir: Hmm. It'd give us an erection.
Estragon: (highly excited) An erection!
Vladimir: With all that follows. Where it falls mandrakes grow. That's why they shriek when you pull them up. Did you not know that?
Estragon: Let's hang ourselves immediately!

Beckett, *Waiting for Godot*, 12

44. "Comme elle [la plante] ils *poussent*, ils *bourgeonnent*, ils *fleurissent*, ils *s'épanouissent*, ils *se fanent*, ils *s'étiolent*, ils *végètent*. Comme elle ils craignent de *prendre racine* à force de rester immobiles, à moins qu'ils souhaitent *s'enraciner* dans quelque paisible domaine rural." Schmidt, *Mandragore*, 5.

45. These include Hasper and Schmitz's *Klein-Erika bei den Wurzelmännchen* (1941), the Tschiedels's *Die Wurzelmännchen* (1949), Larissa, Julius, and van Leeuwen's *Die vergnügten Wurzelmännchen* (1963), Schäfer and Roßdeutscher's *Lustige Wurzelmännchen* (1969), Lazar and Strub's *Akelei und das Wurzelmännchen* (1970), Hummel's *Wurzelmännchen* (1980), and, in the twenty-first century, Kattner and Ehlers's *Wurzelmännchen und seine Freunde* (2006), Struve and Hamann's *Die Hexe Haladusa und das Wurzelmännchen* (2008), and Kraxner's *Ozzibus und die Wurzelmännchen* (2013).

Early European cinema, and German cinema in particular, abounds with mandrake imagery, most prominently as the Germanic legend of Alraune (the feminized version of the noun *Alraun*, or mandrake), which appeared in at least five different German versions between 1918 and 1952.[46] We are far from Machiavelli's comic vision of the mandrake in his play *La mandragola*, published in 1524.[47] In the twentieth century, a deep-seated fear of science drives the new mandrake mythology, which illustrates the horrors of experiments that toy with the biologic constitution of the human by melding the DNA of plants and people. The films seem to suggest that when humanity forgets its place in the natural order, it is destined to be eradicated. The renewed interest in the mandrake was spurred by Hanns Heinz Ewers's 1911 book *Alraune*, about a scientist who conducts an experiment on a woman, impregnating her by means of a mandrake root, which grew on the ground below the gallows where hanged criminals spilled their sperm. The woman conceives a child named Alraune who grows into a perverse and sexually obsessed woman because of her genetic inheritance from the abject plant. Toward the end of the most famous version, Henrick Galeen's expressionist film of 1928, the scientist responsible for the depraved experiment suddenly sees a vision of the mandrake, which morphs through a bright orb of light into the body of Alraune. A similar cinematic strategy is seen in the 1952 version, when a close shot of Alraune's wide-eyed face fades out of focus into the grotesque form of the mandrake. These special effects, possible only through cinema, gave a new and terrifying dynamism to the old legend's tale of transformation. Through the medium of film, the mandrake's horror and ugliness were multiplied when juxtaposed with the beautiful actresses. The films' fears of certain technologies—such as those that manipulate the human's biologic fabric—are underscored by the use of cinematographic technologies that show the bad metamorphosis of humans in a cautionary way, without actually transforming them. A certain strain of Germanic culture could be described as an earth cult, and this obsession in the twentieth century with the mandrake and the Wurzelmännchen confirms

46. These include the 1918 *Alraune, die Henkerstochter, genannt die rote Hanne,* directed by Eugen Illés and Joseph Klein; the 1919 *Alraune und der Golem,* directed by Nils Olaf Chrisander (of which no copy exists); the 1928 *Alraune,* directed by Henrick Galeen; the 1930 *Alraune,* directed by Richard Oswald; and the 1952 *Alraune,* directed by Arthur Maria Rabenalt. A Hungarian version directed by Michael Curtiz and Edmund Fritz also appeared in 1918. The mandrake, as a curative plant, appears in the 2006 film *Pan's Labyrinth,* directed by Guillermo del Toro.

47. The French playwright Jean Vautier reworked Machiavelli's play as *La nouvelle mandragore* in 1952.

a residual paganism and a resistance to modernity, which represents a threat again the natural order of the world.

The root is a foul and base thing, associated with abjection, filth, and taboo. When Samuel Beckett wrote in a letter, "All my trees are down in the cold ground where I shudder to think what is happening to their roots,"[48] he expressed a horror similar to that of Walt Whitman, who, in his poem "This Compost," puts forth the image of plants and soil as revolting processors of death. Bachelard writes that the root is "its own gravedigger," the sum of its activity being, in essence, the burial of itself. Furthermore, the plant's nature is invasive; particularly in the form of weeds, it thrives where there is any bit of soil, even in the cracks of sidewalks or on the roofs of buildings. Roots can drill holes through the most solid human-made structures, turning architectural wonders into ruins.[49] This green invasion is a fundamental fear of humans, who cannot bear the thought of their accomplishments being swallowed by vegetal lawlessness. Georges Bataille has argued that the root invites a moral projection because of its proximity to dirt; it is reasonable to see in the mandrake an evil and unclean miniature of the human.[50] It is snakelike, slithering in the depths like the creature that tempted Eve to eat the apple. Julia Kristeva describes as abject all the unbearable phenomena that blur the border between subject and object; what is unacceptably and revoltingly real forces us to reckon with it. Abjection may take the form of putrescence, filth, the repulsive aspects of reproduction and birth, and other taboos to which the human responds in horror; in Sartre's case, it is the serpentine root that imposes

48. The letter is dated April 7, 1956, and is addressed to Barney Rosset. Beckett, *Letters*, 614.

49. In Saint-Exupéry's *Le petit prince* (1943), the baobab that spreads its roots and has the potential to explode the planet is often read as a figure for National Socialism:

> It seemed that on the little prince's planet, as on all planets, there were good plants and bad plants. Good seeds come from good plants and bad seeds come from bad plants. But seeds are invisible. They remain dormant in the depth of the earth until one of them suddenly decides to wake up. It stretches itself, timidly at first, and then begins to push a charming little sprig inoffensively toward the sun. If it is merely a sprout of radish or a sprig of rosebush, it can be left to grow as it wishes. But if it is a weed, it should be torn out at once as soon as it is recognised. It so happens that there were some terrible seeds on the little prince's planet. . . . They were baobab seeds. The soil of the planet was infested with them. But if you intervene too late, you will never get rid of a baobab. It spreads over the entire planet. Its roots bore clear through it. And if the planet is too small and if there are too many baobabs, the planet explodes.
>
> Saint-Exupéry, *Little Prince,* 25–26

50. See Bataille, *Visions of Excess*, 13–14.

itself on his protagonist in *Nausea*. Whereas the snake is almost never seen in a positive light, the flexible root image fluctuates between benevolence and malevolence. Bachelard recalls Bacon's claim that eating roots, seeds, and other edibles imagined to contain a concentrated vitality can rejuvenate people and lengthen their lives (296). During the long winters, roots store essential nutrients, functioning as the plant's cellar. In some versions of the mandrake tale, it has salvific properties, but these come at a price. Mandrake folklore is one of the best places to look for proof of people's ambivalence toward the botanical underground. While the flower—with its bright colors, clean forms, and appealing fragrance—lends itself to idealization, the monstrous root invites demonization; and yet the former would not be possible without the latter.

Throughout this book, I return again and again to the themes associated with the mandrake. A small, sexually charged version of ourselves, the miniature root person is both good and evil; it embodies the improbable and the inexplicable; and it is a figure of both life and death simultaneously. In this mythology, one can recognize the attempt to understand a difference that separates the plant from the person: when buried, we are either dead already or we die quickly, but this anthropomorphic root thrives in the suffocating soil. If we try to imagine the first human who pulled up a mandrake and recognized in its roots the shape of a person, we might understand the horror and shock of seeing a miniature, buried self that thrives in a state of inhumation. This alter ego, spawned from the seed of malefactors, is paradoxically full of life and able to impart some of its life to the person who finds it. Particularly in Sartre's treatment of the root, we recognize the mandrake legend refashioned for an intellectual and enlightened audience. But Enlightenment does not shine brightly enough to diminish the root's dark and age-old connotations.

*

This introductory chapter has laid the groundwork for those to come. It is essential to understand from the beginning that the root is more than just one metaphor among many. As a figure for the subconscious and for the deep cognitive mechanisms that govern figurative language, the root gives voice to human anxieties about ecological alienation, geographic exile, cultural transmission, and genetic inheritance. The flexible metaphor accommodates both science and mysticism; as will become clear in the following chapters, the tug-of-war between these two ways of thinking has tested the plasticity of the root metaphor. Its most astounding development, analyzed in the final pages of this book, takes shape in the very recent claim that humans and plants share a cellular consciousness, thus annulling the metaphoricity of centuries-old depictions of plants as humans and vice versa. Scholars in philosophy, botany, and anthropology are drawing closer to a conclusion that the idea of a botanical human was never a metaphor. This literalization, made now through scientifically legitimated discourses, closely resembles early paganistic approximations between plant and person. But if the claims are the same, does it matter through which discourse they are made?

In the following chapter, we see how poets have used the root to thematize filiation, environmental estrangement, regional embeddedness, and the process of thinking. Because it welcomes metaphor and has no need to disguise it, poetry is perhaps the most sanctioned and innocuous place for this kind of figurative language. Only when metaphor hides or erases its metaphoricity does it become dangerous. The inherent subjectivity of the poetic voice turns poems into experimental sites where impressionistic thoughts are safely formulated. Perhaps new metaphorizations must be tried out in poetry first before they are allowed to enter the realm of the political.

CHAPTER 2

Radical Poetry

> Writing means having a passion for origins. It means trying to go down to the roots. The roots are always the beginning. Even in death, no doubt, a host of roots form the deepest root bottom. So writing does not mean stopping at the goal, but always going beyond.
>
> EDMOND JABÈS, *The Book of Questions*

One of the most lasting contributions of Bachelard's *oeuvre* was his willingness to take poets seriously, considering literature to be as valid a place as dream analyses or patient testimonies to find evidence of the psyche's inner workings. Bound by few strictures, poetry has tacit permission to access the impressions and intuitions of an attentive subject through imagery and aesthetically conscious language. These impressions, when considered jointly and severally, begin to give a sense of the conscious and subconscious significance of a particular metaphor. Often, a poem is the vanguard of an idea that will take hold in the collective consciousness, creating the conditions under which something new can be said outside literature.

Given the flexibility of the root metaphor, poets commonly turn to it when writing about home, exile, longing, attachment, spirituality, sexuality, nature's will, familial ties, memory, and the underground. For example, the Martinican poet Aimé Césaire's *Notebook of a Return to the Native Land* (*Cahier d'un retour au pays natal*, 1939), is an extended, visceral interrogation of rootedness expressed through the voice of a population pulled between Africa, the Caribbean, and France.[1] Rainer Maria Rilke's *The Book of Hours* (*Das Stunden-Buch*, 1905) uses roots to describe the slow, burgeoning relationship with God. In Antonin Artaud's "Here Lies" ("Ci-Gît," 1947), the family tree is bent into a corrupted closed circuit in which the narrator becomes his own son,

1. The poet writes, "Ma négritude . . . elle plonge dans la chair rouge du sol/elle plonge dans la chair ardente du ciel" and equates himself with the Wandering Jew. Césaire, *Cahier*, 23.

father, and mother. And Yves Bonnefoy uses the imagery of forest undergrowth to convey the notion of primordial memory in his poem "A Voice" ("Une voix," 1958). In these poems and many others, the root becomes the optic through which the problem of connectedness—to homeland, to god, to family, to the past—can be seen and subtly refashioned.

While it would be impossible and unavailing to catalog the innumerable conspicuous uses of the root metaphor in twentieth-century poetry, I've identified four poets whose quintessential application of it push the figure in new directions that give insight into the questions I pose in this book. These poets are Francis Ponge, Paul Celan, Eugène Guillevic, and Denise Levertov. Ponge is more interested in the plant as a whole than roots specifically; but when he does write about them, they are figures for immobility, one of the primary distinctions between the human and the plant. Root imagery recurs prominently in Paul Celan's poems and, in the most salient appearances, is a figure for absolute negation, of a family lineage and of a people. Eugène Guillevic, in his celebration of regional rootedness, uses the root as an occasion for launching thought experiments about time, death, the subterranean world, and vegetal will. Denise Levertov, a British-born American poet, did not publish poetry in French or German, but she was heavily influenced by both Celan and Guillevic. I see in Levertov's work a transformation of these two poets' renderings of the root into a figure for ecological alienation.

PONGE AND THE PLANT'S IMMOBILITY

Bachelard valued in particular the poetry of Francis Ponge (1899–1998), who showed his readers that "by working oneirically at the interior of things, we go to the dreaming root of words."[2] Ponge's poetry is perhaps the most botanically oriented in the twentieth-century French literary canon. The language of vegetal proliferation is found throughout his corpus, particularly in his *Le parti pris des choses*,[3] *La rage de l'expression, Le pré*, and *La fabrique du pré*. He notes the comparison—made frequently in literature and philosophy—between the proliferation of an idea and of a plant; he likens his thoughts

2. "Comme l'indique clairement Francis Ponge, en travaillant oniriquement à l'intérieur des choses, nous allons à la racine rêveuse des mots." Bachelard, *Terre*, 11.

3. These include, for example, "Rhum des fougères," "Les mûres," "L'orange," "Les arbres se défont à l'intérieur d'une sphère de brouillard," "La mousse," "Faune et flore," and "Végétation."

on a particular day to a eucalyptus.[4] However, these ideas are headless and centerless. Michael Marder is attracted to Ponge's vision of plantlike thinking that begins in the middle because it resists "the idealist insistence on the spirituality of the blossom and the materialist privileging of the root."[5] Reading Ponge through the ideas of Deleuze and Guattari, Marder considers all plant growth rhizomatic, since it begins *in medias res*. Ponge describes plants as "a kind of three-dimensional tapestry"[6] and as "living crystals,"[7] an image that calls to mind Henri Atlan's well-known work *Le cristal et la fumée: Essai sur l'organisation du vivant*.[8] Plants are living reminders of the way life can organize itself so differently from its organization in our own bodies.

Regarding rootedness, Ponge becomes extremely slippery, alternating at times between a celebration and a condemnation of radical immobility. He notes a primary difference between the animal body and the vegetal body: "Fauna moves, whereas flora unfolds before the eye."[9] In its false immobility—for the plant is not actually still; it creeps and spreads itself in far more creative ways than the human—the plant is seen as a passive being that unfolds *for* the eye, as if the human's perception of it were its only purpose. Roots may provide stability for the plant, but they are by no means anchors, despite the tendency to metaphorize them as such. Ponge selects this common metaphor

4. Ma *pensée*, aujourd'hui, *comme* L'EUCALYPTUS, est un arbre fier,
mais de bois assez friable,
de pose et de geste majestueux;
original;
quelque peu étranger parmi les autres arbres.

Ponge, *Nouveau recueil,* 116

In *Les déracinés,* Maurice Barrès ventriloquizes Hippolyte Taine, who says while looking at a tree, "Sentez-vous sa biographie? . . . Quelle bonne leçon de rhétorique, et non seulement de l'art du lettré, mais aussi quel guide pour penser!" Barrès, *Déracinés,* 215.

5. Marder, *Plant-Thinking*, 63.

6. Ponge, *Tome premier*, 102. Jung describes how plants seemed to him the visible artisanal handiwork of God: "The earthly manifestations of 'God's world' began with the realm of plants, as a kind of direct communication from it. It was as though one were peering over the shoulder of the Creator, who, thinking Himself unobserved, was making toys and decorations." Jung, *Earth,* 28.

7. "Les végétaux sont des cristaux vivants." Ponge, *Nouveau recueil*, 104.

8. Atlan argues that life tends to organize itself along a spectrum, with, at one end, the perfect symmetry found in some crystals and, on the other end, the "evanescent forms of smoke." Atlan, *Entre le cristal*, 5.

9. "La faune bouge, tandis que la flore se déplie à l'œil." Ponge, *Tome premier*, 90.

in his "L'opinion changée quant aux fleurs," transforming the plant into an anchored nautical ship:

> In the plant . . . this knot could be found in a certain place, between the stem and the root (not far from its exit from the earth) where growth happens *in both directions* (upward and downward); center or knot conserving the memory of the seed; all in all, at the center of gravity: at the base of the mast and at the top of the keel.
>
> In the spirit of this new metaphor (that of the plant-ship), one could also imagine the roots as anchors, as a system of anchors. And one could say that the plant is anchored organic material (having dropped anchor).[10]

For Ponge, the plant is ultimately a constrained entity, rooted absolutely in place.[11] He describes plants as voiceless, paralyzed, and composed of pure *excroissance*, or outgrowth. In his "Faune et flore" from *Le parti pris des choses*, the plant is condemned to repetition, reiterating "a million times the same expression, the same leaf."[12] It purges "green vomit" in the form of foliage.[13] This abject vision at first seems to indicate some intense botanophobia. Fear of the vegetal world is common in the literature of the late nineteenth and twentieth centuries; as I show in chapter 5, Sartre's fear of the tree makes him adopt it as a figure for the intrusiveness of Being. However, the physical abjection Ponge describes in the plant is soon replaced by a beautiful abstraction.

10. Chez le végétal . . . ce nœud pourrait-il se trouver à un certain endroit, entre tige et racine (pas très loin de la sortie de terre), où se produit la poussée *dans les deux sens* (vers le haut et vers le bas); centre ou nœud conservant le souvenir de la graine; somme toute, au centre de gravité: à la base du *mât* et au sommet de la *quille.* /Dans l'esprit de cette nouvelle métaphore (celle du végétal-navire), on pourrait aussi concevoir les racines comme des ancres, comme un système d'ancres. Et dire que le végétal, c'est la matière organique à l'encre (ayant jeté l'ancre).

Ponge, *Nouveau nouveau recueil,* 106

Roots display what is called "positive gravitropism" since they obey earth's gravitational pull toward the center; the plants' stems show negative gravitropism, since they grow against gravity.

11. "Le temps des végétaux: ils semblent toujours figés, immobiles." Ponge, *Tome premier*, 93.

12. Ponge, *Tome premier*, 91. This idea of repetition and reiteration mirrors Paul Valéry's idea that "knowledge extends like a tree, by a process identical to itself; by repeating itself. *Novat reiterando*. [It renews itself through repetition.]" ("La connaissance s'étend comme un arbre, par un procédé identique à lui-même; en se répétant. Novat reiterando.") Valéry, *Cahiers,* 3:273.

13. "[Les plantes] laissent échapper un flot, un vomissement de vert." Ponge, *Tome premier*, 91.

Its proliferation expresses an enthusiasm for life; as it spreads, the plant exclaims, "Excelsior! Longior!"[14] He sees in the plant a figure for writing; in contrast, animals embody oral communication. The plant and the writer are laborers who, although immobile, glean from around them all that they need for their labor. The poet becomes emphatic, switches to all capital letters, and appeals loudly to the reader: "HENCE THE ESSENTIAL QUALITY OF THIS BEING, FREE FROM ALL DOMESTIC AND DIETARY WORRIES ALIKE BY THE PRESENCE OF AN INFINITE RESOURCE OF FOOD AROUND IT: *Immobility*."[15] The generalized presence of stationary plants across the planet makes all humans look like "vagabonds, homeless, with neither hearth nor home."[16] This relativization of mobility works to undo the source of contention between the *nomades* and the *sédentaires* by positing that in relation to plant life, we are all wanderers. Rootedness makes it impossible for the plants to make mistakes, lose their bearings, or hesitate, giving them an advantage over the animal and the human. Alternating between a laudatory and a vilifying conception of rootedness, Ponge's ambivalence participates in his standard process-oriented, essayistic writing practice, which tries out as many possibilities as conceivable. The "opinion changée" is part of his writerly method. Because of its immobility, the plant is a living object of study for the human. It cannot run from our gaze like the animal that scuttles away in the underbrush. Modeling an embeddedness in the environment that the human will never achieve until death, the plant does not go in search of the world; the world comes to it. Ponge's lyrical rendering of this immobility oscillates between plant as imprisoned and plant as luxuriantly lodged, the most contextualized of all living entities. Ponge tries customarily to side with things, but one wonders whether he considers the plant a thing or a being. Science tends to treat it as the former, poetry the latter. Ponge's scientifically inflected poetry tries to see it as both, sometimes simultaneously. As I show in the final pages of this book, science and poetry seem to have found a

14. "Et soudain, voici qu'à partir des mêmes points, en tous sens, dans toutes les directions, vers tous les azimuts [la plante] se rélance,—toujours plus loin et plus haut! Excelsior! Longior!" Ponge, *Nouveau nouveau recueil*, 128.

15. "D'OÙ LA QUALITÉ ESSENTIELLE DE CET ÊTRE, LIBÉRÉ À LA FOIS DE TOUS SOUCIS DOMICILIAIRES ET ALIMENTAIRES PAR LA PRÉSENCE À SON ENTOUR D'UNE RESSOURCE INFINIE D'ALIMENTS: *L'immobilité*." Ponge, *Tome premier*, 97.

16. Ponge, *Nouveau recueil*, 105. In Saint-Exupéry's *Le petit prince*, the prince stops to ask a flower if it has seen any men passing by, to which the flower replies, "The wind blows them around. They have no roots, which makes their life rather trying." Saint-Exupéry, *Little Prince*, 71.

new synthesis in recent theories that use science and philosophy to defend the poetic view of the plant's ontology.

INTO THIN AIR: CELAN'S "RADIX, MATRIX"

We might say that Ponge is the plant's poet; but the root has its own set of verse-makers. Paul Celan (1920–1970), the Jewish German-language poet from what was the Kingdom of Romania, today Ukraine, spent the decades between the murder of his parents Fritzi Schrager and Leo Antschel in the camps and his own suicide by drowning in the Seine attempting to say what language cannot. Celan's fluency in Romanian, German, French, English, and Russian, and his strong competency in Hebrew and Yiddish, meant that he had at his disposal a breadth of vocabulary, grammar, and language-specific turns of phrase; this did not make easier the task of putting language to what was but is no longer. Vivian Liska has analyzed the transformation of the root metaphor in Celan's corpus, claiming that in its early iterations, the metaphor creates a link between "poetry, blood, and soil," which then becomes heavily encrypted in his later work.[17] Unlike many Jewish writers and thinkers, Celan never adopts the "negative myth of origin" (56) of thriving in uprootedness, in a move that Vilém Flusser has called "taking up residence in homelessness."[18] Instead, Liska argues, "By excavating its own ground, Celan's language uproots itself" (56). His neologisms and disorienting hyphenations, syntax, and rhetorical figures made his late poems some of the most difficult in the European canon.

Consider "Radix, Matrix," published in his collection *Die Niemandsrose* (*No One's Rose*) in 1963. This poem imparts the nonavailability of a long series of futures, personal and collective, which take nonform as the nullified vision of what might have been a family tree.[19] John Felstiner attributes Celan's renewed interest in Judaism around the time the poem was written to his readings of Osip Mandelstam, Martin Buber, Gershom Scholem, and other Jewish thinkers.[20] The Judeocide recounted in the poem is both the specific, maternal

17. Liska, " 'Roots against Heaven,' " 47.

18. See Flusser, "Taking Up Residence."

19. The same title was given to a collection of essays by and about the Polish architect Daniel Libeskind, who designed the Jewish Museum in Berlin and who was one of the designers of the One World Trade Center building. For more on the connection between Celan's and Libeskind's aesthetics, see Eric Kligerman's "Ghostly Demarcations: Translating Paul Celan's Poetics into Daniel Libeskind's Jewish Museum in Berlin," *Germanic Review* 80, no. 1 (Winter 2005): 28–49; and his book *Sites of the Uncanny*.

20. Felstiner, "Mother Tongue," 123.

one and the generalized European massacre. Key words in the poem—*Heimat, Wurzel, Matrix*—belong to Bachelard's catalog of images of terrestrial repose. One of the first things a reader may notice about "Radix, Matrix" is the Latinate double X in the title. X marks the spot, that place in the ground where the person and the race originated. In genetic terms, the double X is the chromosomal signature of the female. The poem's femininity, or rather, its maternity, is signaled in the title both by the presence of the double X and by the matricial second term. Of the word "radix," Isidore de Sevilla claimed in his *Etymologiae* that it derives from the spokelike *radii* of the root's form.[21] To imagine a root as a wheel just below the soil's surface, a radial form whose lines depart from a center, is to put in conversation the collective symbolic associations of root and wheel. The root is a figure of stasis, the wheel of movement. Rootedness invokes the past; the wheel carries us toward the future, toward what Fortuna will grant or thrust upon us. The spoked *radix* is a botanical Wheel of Fortune. The radius is also a ray; imagine lines of light emitted from a center, a solar figure, whose form resembles the root's reaching extremities.[22] Celan's root is a blacked-out sun.

The first stanza of this free-verse poem cuts words at their prefixes. The "Ver-" of "Verschwisterte," the "Zu-" of "Zugeschleuderte," and the "Be-" of "Begegnete" are uncoupled from their words, beginnings slit from their ends. In Celan's first stanza, the *trait d'union* becomes a *trait de séparation*. Because this poem is about negation, lessening, lowering, and extrication, it is worth mentioning that the hyphen, of which there are nine in the poem, is also the mathematical sign for subtraction. These deverbal adjectives derived from prepositional verbs (*verschwistern, zuschleudern, begegnen*) turn actions into descriptions, creating the impression of induced passivity.[23] The other hyphenated formulations are the strange pairing *Aber-Nacht* (literally "but-night") and *Aber-Du* (literally "but-you," in the informal second person), an equation that creates equivalency between "you" and "night." The *aber*, typically a coordinating conjunction, is grammatically unifying in that it binds two equal clauses, yet adversative in its marking of a contrast between the two.

21. "The root (*radix*) is so called because fixed in the ground in the manner of 'radiating spokes' (*radius*), as it were, it goes down deep, for natural philosophers say that the depth of the roots is equal to the height of the trees." Isidore, *Etymologies*, 341.

22. "Others say roots are named for their similarity to rays (*radius*), or because, if they are cut away (*eradere*), they do not grow back." Ibid.

23. Two other deverbal adjectives, *gemordete* and *stehende*, appear in the third stanza.

The use of *du*, which some read as an address to the mother, has also been taken as a sign of the fusion between the poem and the address. In Philippe Lacoue-Labarthe's reading, "the poem melds with the address itself; there exists only a sort of nomination without a name, a 'saying-you.'"[24] Werner Hamacher sees in its address a poem that makes a specific appeal: "In asking after *radix, matrix*, the poem asks after its own ground. . . . It no longer speaks the language of a race or a sex, which could be the ground, center, origin, father, and mother; it speaks—deracinated, dematricated—the language of the murdered *Geschlecht*."[25] The appeal is made across an impossibly wide chasm whose blank breadth includes both space and time. The poem's emphasis on distance comes through in the second stanza, both as a spatial (*da*, or "there") and a temporal remoteness (*damals*, or "back then"). In fact, the "d" alliteration (*Damals, da, da, damals, da, du, den*) and the "a" assonance (*Acker, abschrittst, allein*) that mark the stanza spell *da*, the inversed first letters of *Aber-du*. John Felstiner sees in this passage a distant memory of the poet's mother, expressed in a language that "verges on infant dada or stammering."[26] But it is also one of the many gestures of distancing, like the high-contrast silhouetting of the murdered *Geschlecht* against the sky, the impossible communication between human and stone ("Wie man zum Stein spricht"), and the vocabulary of temporal and spatial separation (*vorzeiten, da, dorthin, Hinab*). Readers may also notice the four colons, used typically to provide examples or to introduce an expansion or illustration of the preceding clause, which here create the effect of falling deeper into an example of an example, a *mise-en-abîme* downward toward the *Hinab*, the "down there" of the last stanza.

The sexual nature of the third stanza is extremely difficult to render in languages other than German, as Pierre Joris explained regarding his English translation of the poem. He notes that the word *Geschlecht* "carries a constellation of meanings: sex, gender, race, generation, family, lineage, species, genre."[27] While Joris translated "Rute und Hode" as the neutral "Rod and ball," which does not necessary trigger an explicit sexual image in the Anglophone reader's mind, he acknowledges in the introduction the sexual nature of the terms and draws a further connection between the testicle and the witness,

24. Lacoue-Labarthe, *Poetry as Experience*, 83.
25. Hamacher, "Second of Inversion," 368.
26. Felstiner, *Paul Celan*, 176.
27. Joris, introduction to Celan, *Paul Celan*, 28.

through the Latin root *testis*.[28] Hamacher argues that the *Rute* and the *Hode* are rather figures for a coupling, not simply the components of male genitalia:

> Rod and testicle isolate the sexual aspect of *Geschlecht* but not, as it might appear, its phallic aspect. For *Rute* is the word for the *radix*, which stands in Latin not only for vegetable root, for origin, source, firm ground, and soil, but also, as in *radix virilis* for the masculine member. *Hode*, on the other hand, derives from the Latin *cunnus*, the pudenda, and so corresponds to *matrix*, which just as much as *radix* means source, origin, and stem, but in its feminine aspect as progenitrix, womb, uterus. Only together, as *Rute* and *Hode*, and brought still closer in the asyndeton of the title, as *radix, matrix*: as the coupling of the masculine and feminine sexes are they "that *Geschlecht*, that murdered one, the one/standing black in the sky." Only in this coupling do they fulfill the figure of immanent inversion that the erected abyss presents.[29]

The strange negative construction of the "erected abyss" depicts absence as a buildable phenomenon.[30] But shadows have no density; the erection is only a barren profile, the silhouette of nothing. John Felstiner has argued that there is a triple pun in Celan's choice of *Rute*, including one on the name Ruth, the grandmother of the biblical Jesse, father of the Israelites' king David: "Since *Rute* means 'rod' and 'penis,' and in German the name Ruth is pronounced like the English 'root,' a triple play between languages unites—without at all reconciling—the murdered mother, the rod of miracle or anger, and a radically threatened people."[31] Martine Broda translates *Rute* into *verge* in French, which means the man's generating mechanism, derived from the Latin *virga*, or branch. The poem abounds with the language of genetic filiation (*Matrix,*

28. Joris explores Celan's use of the verb *zeugen* in his corpus: "The German word *zeugen* also has the meaning 'to beget, to generate,' a meaning kept more or less alive in the English word *testify* via its Latin root *testis,* which refers both to the 'witness' and to 'testicle' (as the 'witness' of virility). One also has to keep in mind the 'Rod and ball,' the 'Hode' from the poem 'Radix, Matrix.' (Another semantic extension would lead us to 'testament,' 'testamentary'—clearly terms that can also play into the witness complex)." Joris, introduction to Celan, *Paul Celan,* 32. Derrida argues that Valéry's Monsieur Teste is so named not because of the *tête,* or head, but because of his role as *testis,* witness. See Derrida, "Les sources de Valéry," 593.

29. Hamacher, "Second of Inversion," 367.

30. Daniel Libeskind wrote in his book *Radix-Matrix*, "We are witness to the events in which the architecture of presence turns into the architecture of absence." Libeskind, *Radix-Matrix,* 141.

31. Felstiner, *Paul Celan*, 177.

verschwistern,[32] *Geschlecht, Rute, Hode*). A word that is not etymologically related but phonically similar in French is *verger*, the orchard, whose etymology leads back to *viridarium* and *viridis*, or green. The poem's vocabulary is heavily botanical: *Acker* (field), *Wurzel* (root), *Fruchtboden* (fertile soil), and *wildblühenden* (wild-blooming). Fertility and greenness, *verge* and *verger*, become emptied-out ideals. In the closing stanza, Celan's term *Fruchtboden*[33] conflates fruit and ground, the necessary pairing for germination to happen. All of the empty figures (*Abgrund*, *Nichts*, "o") in "Radix, Matrix" replace the productive somethingness of fertility with naught. Derrida sees in the poem the emblematic circumcisive cutting he identifies as a feature of Celan's poetics,[34] but the cutting is also the neutering sort. The black silhouette of the *Geschlecht*, standing alone in high contrast against the sky, shows it to be the castrated shadow of its living self.[35] The desexing of the sex culminates as an empty profile against the heavens. *Radix, Matrix* is a de-seeded poem.

The *radix* of the title returns in the fourth stanza, de-Latinated and Germanized:

> (Wurzel.
> Wurzel Abrahams. Wurzel Jesse. Niemandes

32. Felstiner translates *verschwisterte* as "siblinged" (ibid., 176), and Joris translates it as "dissister." Joris, introduction to Celan, *Paul Celan,* 83.

33. *Fruchtboden* is translated by Joris merely as "receptacle," by Broda as "sol à fruits," and by Kligerman as "fertile soil." Kligerman, *Sites of the Uncanny,* 248.

34. Derrida comments, in *Sovereignties in Question,* 55:

> If all poets are Jews, they are all, the poets, circumcised or circumcisers. This opens up, in Celan's text, a tropic of circumcision that turns from ciphered sores toward reading-wounds, all cut words, notably in "Engführung," where a thread can be followed that passes through "points of suture," closed up tears or scars, words to be cut off that were not cut off, membranes stitched back together, and so on. . . . The evocation of the exterminated race designates the race and root of no one: black erection in the sky, verge and testicle, race and root of no one. Uprooting of the race, but equally so of the sex (*Geschlecht*) in "Radix, Matrix."

35. Compare Celan's stanza with this one from Jacques Prévert's *Arbres*:

> Tout seul un olivier
> jette désespérément
> vers le ciel calciné
> deux bras carbonisés
> comme un nègre lynché
>
> Prévert, *Arbres,* 43

Wurzel—o
unser.)

([Root.
Root of Abraham. Root of Jesse. No one's
root—o
ours.])

Joris makes an effective pun when he describes this parenthetical segment of the poem as a "*parent thesis*."[36] Concentrated and cordoned off from the rest of the poem by these curved brackets, the exhausted roots are placeholders in a spot usually reserved for explanations or afterthoughts. The compressed genealogy presented here ends with no one. The anaphoric *Wurzel/ Wurzel/Wurzel*, repeated in the same fashion as the anaphoric *du/du/du* in the first stanza, creates an analogy between them. If "du" is the mother, the radix-matrix binary is reaffirmed through these anaphora. The Jewish patriarch Abraham and the progenitor Jesse, father of David, are germinative figures who gave an original form to the great family tree. In a letter from his friend Nelly Sachs regarding Celan's translation of Valéry's *La jeune parque*, she wrote, "You have touched the roots of language as Abraham did the roots of faith."[37] In many visual renderings of Jesse, he reclines as the tree of multiplied futures sprouts from his body. This recumbent burgeoning emphasizes his role as originator, with the tree being an ideal figure to convey the Latin *oriri*, which means to rise or become visible, from which "origin" derives. As shown in Ponge's formulation, the tree unfolds itself as a spectacle for the eyes, going about its forking proliferation as a singular entity or a forest, multiplying its presence. Just as the rose in the collection's title is no one's rose, the root is no one's ("Niemandes/ Wurzel"). The "we"—descendants of these fathers, Abraham and Jesse—will not make it outside the brackets alive. All that is ours is "o." Broda, in her French

36. Joris, introduction to Celan, *Paul Celan*, 28.

37. Celan and Sachs, *Correspondence*, 18. ("Sie haben die Wurzel der Sprache gefaßt wie Abraham die Wurzel des Glaubens." Celan and Sachs, *Briefwechsel*, 31.) Sachs used a similar formulation in one of her poems: "Nun hat Abraham die Wurzel der Winde gefaßt/denn heimkehren wird Israel aus der Zerstreuung" (Now Abraham has grasped the root of the winds/ because Israel will return home from the dispersal). In many of her poems, she emphasizes the edaphic or soil-boundness of the human, with poems that begin with lines like "Wie leicht wird Erde sein," "Völker der Erde," "Erde, Planetengreis," "So steigt der Berg," and "Wer von der Erde kommt."

translation, renders it "o" rather than "ô," which accommodates much more effectively readings, like Hamacher's, that take the "o" not as simply an irrepressible interjection of sorrow and a lamentation for the hole where the root once was but as the absolute void, the absence of even the ground in which to make a hole.[38] Celan used in many of his poems the idea of a vanishing act, a disappearance into thin air, appropriating the derisive *Luftmenschen* label and lyricizing it.[39] His "Todesfuge" replaces an earthen grave with a grave in the clouds ("ein Grab in den Wolken") or a grave in the air ("ein Grab in der Luft"), images thought to be inspired by the poem "Er," written by a former classmate named Immanuel Weißglas, which begins, "Wir heben Gräber in die Luft" ("We raise graves in the air"). The first line of the last poem in Celan's *Die Niemandsrose* reads: "In der Luft, da bleibt deine Wurzel, da, in der Luft" ("In the air, there stays your root, there, in the air").[40] The palindromic structure of this line, which follows the same structural logic backward and forward, leaves the root—your root—suspended permanently there in the middle of the phrase. The groundlessness of the new grave parallels the lived groundlessness of a people without a place, suspended between nations. Nicolas Berg has argued that the history of the *Luftmenschen* metaphor displays a "tension between an anthropologizing essentialism and a literary irony" but that the

38. Hamacher writes, "The terse genealogical catalogue, which contains an infinite genealogical promise, is—after 'Jesse'—no longer subject to a chiasmus alone, to an inversion of the syntactical elements that sets 'no one' in the place of 'root,' but this chiasmus is itself subjected by the line-break to an interruption that opens in the place of 'root' an empty place—a 'pause,' a 'hiatus,' a 'lacuna.'" "Second of Inversion," 369. In the note to this citation, Hamacher writes that "the 'o'—placeholder for the 'Wurzel'—is to be read not only as a plaintive sound but also as the graphic sign of the annulment of the root and of ours" (370n25). It is worth noting that in his introduction to *Paul Celan*, Joris writes "o" while in the full translation, he uses "oh," probably a lapse.

39. George Steiner wrote, "Hitler spoke derisively of *Luftmenschen*, of the Jew as an unhoused 'creature of the air.' But the air can be a realm of freedom and of light. 'Be a fertilizer among men,' urged one of the founders of Israel, 'compacted into one nation you may become dung.' Nationalism, of which Israel is necessarily emblematic, tribal ingathering, seems to me not only foreign to the inward genius of Judaism and the enigma of its survival. It violates the imperative of the Baal Shem Tov, master of Hassidism: 'The truth is always in exile.'" Steiner, *My Unwritten Books*, 122. Rochelle Tobias writes of this image in Celan's poetry: "The phrase 'floating ones' recalls the fate of eastern European Jews, who were deported in large numbers to the camps and whose only grave often was 'in the air' above smokestacks." Tobias, *Discourse of Nature*, 68.

40. Celan, *Gedichte*, 290.

"term itself says little about Jews."[41] He illustrates how the word *air*, which often has positive connotations of freedom, was transformed into a code word for detachment, danger, homelessness, and other threatening associations.[42] Using an elemental logic, the Germans saw themselves as terrestrial stewards, while the Jews were dismissed as mere airy transients. The zero-root, a kind of botanical nothingness and interrupted fertility, is nonetheless permanent. It remains a *kleines Wurzelgetraüm*, the small root dream, that makes death the poet's dwelling.[43] Celan's root vision is inflected with absence, matricide, and degenerated genesis, reminding us of the old unproven speculation that *race* and *radix* are etymologically connected. The root is always in proximity to death; it lives through it.

GUILLEVIC'S RADICAL TRYING

While Francis Ponge is often praised as France's most botanically minded twentieth-century poet, Eugène Guillevic (1907–1997) may be the most root-minded. A Leftist, he believed that everyone was "either on the side of the oppressed or on the side of the oppressors,"[44] although he regretted his early support of Stalinism before he understood the dictatorial extremism of the Soviet Communist Party. As a trained economist, he spent much of his life researching how fiscal politics touches the lives of real people. He was

41. "An vielen der Beispiele, die dieser Essay diskutierte, konnte verdeutlicht werden, dass die Spannung zwischen einem anthropologisierenden Essentialismus und einer literarischen Ironisierung den Begriff geprägt hat und ein wesentliches Moment seiner umkämpften Semantik darstellt. Der Begriff selbst sagt wenig über Juden aus." Berg, *Luftmenschen*, 209.

42. "Losgelöstsein, Verlorenheit, Unsicherheit, fehlender Grund und Boden sowie der Verlust von Heimat sind die wesentlichen Felder seiner assoziativen Bestimmung." Ibid., 22.

43. This expression comes from one of Celan's later poems:

> Kleines Wurzelgetraüm, das mich hält,
> blutunterwaschen,
> keinem mehr sichtbar,
> Todesbesitz
> (Little root-dreamgrowth, which holds me
> washed with blood,
> no longer visible to anyone,
> possession of death)

Celan in Liska, "'Roots against Heaven,'" 56.

44. "On est ou avec les opprimés ou avec les opprimants." Guillevic, *Un brin d'herbe*, 25.

close with Paul Eluard, a fellow Communist, and was one of France's more politically engaged poets of the mid-twentieth century. Guillevic—who signed his work only with his surname, refusing the name given him by his oppressive mother—called poetry his "source" and his "roots."[45] A devotee of dictionaries and etymologies, Guillevic rarely let a word slip by without tracing its most fundamental roots. He described himself as a "man of prehistory" who "lives in what is elementary."[46] He resisted nationalism in favor of regionalism.[47]

Guillevic is best known as the poetic celebrant of Breton regionalism. In his obituary in the *Independent*, the necrographer James Kirkup writes of Guillevic, "The poet had a sturdy figure that made him resemble a walking menhir from his native Brittany."[48] Kirkup's allusion to the prehistorically erected stones that dot the Breton countryside, which Guillevic celebrates in his poetry, is standard fare in biographical depictions of the poet. Perhaps his self-fashioning through the countless filmed and printed interviews he conducted throughout his lifetime helped create a cohesive—and often clichéd—depiction of a man sprung up directly from a land of Celtic ancestors. We find menhirs on the cover of Denise Levertov's translation of his poems, which presents him this way in the introduction:

> Guillevic is a Breton, born at Carnac of peasant stock. His poetry has deep roots in that inheritance. The great ritual places of the Celts, whether in Wales or Cornwall, Ireland or Brittany, the places where the great and small stones or *menhirs*, are gathered in powerful and enigmatic testimony of forgotten certainties, are landscapes of a profound austerity. In such landscapes the senses are undistracted from the elemental: rock, sky, sea are there not backgrounds but presences. Beginning in such a place, Guillevic learned to recognize all else in life likewise as *presence*, not as incidental properties. Man, bird, cloud, lake, night, death, the sunlight—he knows them as Powers and Principalities, meets

45. "I began to write poetry again that was really mine, like the kind I used to write. Once again I found my source, my roots." Guillevic, *Living in Poetry*, 111.

46. "Je vis dans la préhistoire. J'ai coutume de dire que je suis un homme de la préhistoire. Je vis dans l'élémentaire, oui, mais je vis quand même dans le monde moderne." Guillevic, *Un brin d'herbe*, 59. In his *Terraqué,* he writes, "Sans fin la forêt creuse la terre sèche/Où la préhistoire est restée couchée." Guillevic, *Terraqué,* 131.

47. "Je ne suis pas chauvin français puisque je suis breton." Guillevic, *Un brin d'herbe*, 70. Regionalism is where he stops, claiming that the sharper specificity of *départementalisme* is too extreme (15).

48. Kirkup, "Obituary."

them face to face, and disdains the folly of attempting to use them as mere autobiographical adornments. This atheist is a radically religious poet.[49]

I cite this passage at length to give another clear example of radical portraiture in its most explicit form, as described in chapter 1 in the introductory passage of Maupassant's *Le horla*. The filial logic implied in Levertov's statement that he was born "of peasant stock" might strike us as outmoded today; all talk of stock, breed, lineage, pedigree, and parentage is used now more often to talk about animals than people. As we see in Levertov's passage, the features of the land mark the person. Guillevic is the poet that he is only because of the countless indiscernible ways in which the land left its signature on his psyche. This belief, analyzed throughout this book, relies on a specific philosophy of the human, one that takes the person as a creature whose immanent domain is the ground up from which it raises itself vertically toward the heavens, like a plant. This need to believe that people somehow belong to the land that yielded them, thus property of their property, is one of the common ways people make sense of the seeming arbitrariness of their birthplaces, their *incunabula*. The human, made of and sprung from the *humus*, is first swaddled in soil. Transcendence is imaginable, as I show in chapter 3, only as the result of some form of immanent rootedness, most often in a place. Note Levertov's use of the word "radically"—a problematic term, as we'll see later—to describe the religiosity of Guillevic's atheism. Radical portraiture always uses a language of original autochthony, of the geographically lesioned body and mind. The interesting part is that Levertov herself never felt rooted anywhere; perhaps it is for this reason that she was attracted to the poems of a rooted son.[50]

In his literature and interviews, Guillevic insists on the placeness of his poetry; nearly everything he wrote is topophilic.[51] He surrounded himself with other earthen poets—like Kenneth White, the Scottish founder of geopoetics—equally invested in the relationship between place and person.[52] From an

49. Levertov in Guillevic, *Guillevic*, vii.

50. She wrote in 1983, "I've had to manage without the deep sense of roots in a particular town or countryside which is so valuable to many writers." Levertov, *New and Selected Essays*, 60.

51. W. H. Auden coined the term "topophilia" in his introduction to a collection of John Betjeman's poetry called *Slick but Not Streamlined* (1947). Gaston Bachelard also uses the term in his *Poetics of Space.*

52. The titles of his friend Jean Follain's books evince a terrestrial obsession: *Chef-Lieu, Chants terrestres* (1937), *Ici-Bas* (1941), *Territoires* (1953), *Palais souterrain* (1953), *Notre monde*

elemental perspective, Gaston Bachelard considered him a terrestrial poet.[53] Guillevic published his first book of poetry, *Terraqué* (which means "terraqueous"), in 1942, the same publication year as Ponge's *Le parti pris des choses*. Having spent part of his childhood in Alsace, Guillevic was a fluent reader of German and kept Georg Trakl's poetry at his bedside during adolescence.[54] He was also greatly influenced by Christian Morgenstern, who was best known for his absurdist poetry and his later commitment to anthroposophy. Consider, for example, Morgenstern's poem "The Two Roots" ("Die zwei Wurzeln") published in his collection *Gallows Songs* (*Galgenlieder*, 1905), a copy of which Guillevic owned:

> Zwei Tannenwurzeln groß und alt
> unterhalten sich im Wald.
>
> Was droben in den Wipfeln rauscht,
> das wird hier unten ausgetauscht.
>
> Ein altes Eichhorn sitzt dabei
> und strickt wohl Strümpfe für die zwei.
>
> Die eine sagt: knig. Die andre sagt: knag.
> Das ist genug für einen Tag.
>
> [A pair of pine roots, old and dark,
> make conversation in the park.
>
> The whispers where the top leaves grow
> are echoed in the roots below.
>
> An agèd squirrel sitting there
> is knitting stockings for the pair.

(1960), *Appareil de la terre* (1962), *Cheminements* (1964), *Célébration de la pomme de terre* (1966), and *Ordre terrestre* (1986).

53. Guillevic and Bachelard shared an interest in each other's work. In an interview, Guillevic states, "Bachelard put me in the category of men of the earth" ("Bachelard me rangeait dans la catégorie des hommes de la terre"). Guillevic and Jean, *Choses parlées*, 22.

54. "Trakl a été mon livre de chevet pendant toute la fin de mon adolescence, il a été pour moi très important." Guillevic, *Un brin d'herbe*, 20. He cites other important German influences, including Rilke, Novalis, Mörike, Goethe, Hölderlin, and Brecht.

The one says: squeak. The other: squawk.
That is enough for one day's talk.][55]

This poem, like a child's nursery rhyme, includes anthropomorphized protagonists whose circumstances in the grand scheme of things are very small; no epic themes or philosophical heaviness here. But this humor was of great appeal to Guillevic, whose own simple style and pared-down poetics maintain at all times a reliable lightness. What he appreciated most about Morgenstern was his frequent untranslatability, something he confirmed firsthand when trying to translate the German poems into French. Another possible source of appreciation, perhaps a subconscious one, is Morgenstern's connection with anthroposophy, the philosophy founded by Rudolf Steiner, who describes it this way: "Anthroposophy has its roots in the perceptions—already gained—into the spiritual world. Yet these are no more than its roots. The branches, leaves, blossoms, and fruits of anthroposophy grow into all the fields of human life and action."[56] Steiner considered Morgenstern a true representative of anthroposophy, and the language of roots is omnipresent in his anthroposophic poetry and epigrams. In a later chapter, I show the heavy reliance on root metaphors in anthroposophic circles, but I mention this influence here because of Guillevic's own special attraction to roots and the importance he placed in his poetry on the cultivation of a deliberate autochthony. Guillevic knew little of Morgenstern's life, saying only that he'd heard that he died a theosophist,[57] but the features of anthroposophical thought—such as the desire to wholly integrate the human into work, education, domestic life, and nature—can be found in most of Guillevic's poetry. The terrestrial sympathies of Guillevic, Morgenstern, and the anthroposophists are common in regionalist, domophilic (home-loving), and ecocentric literature.

Guillevic published an entire poetry book on the theme of roots in 1973, entitled *Racines* and accompanied by "compositions" made by Robert Blanchet, which included imprints made by actual roots. Many of the ideas in this extended poem seem to come directly from Bachelard's chapter on roots, a

55. Morgenstern, *Gallows Songs*, 118–19 (the German is on p. 118 and the translation on p. 119). In a letter dated October 26, 1908, to his wife Margareta, he wrote: "Du und ich wie ein Baum aus zwei Wurzeln" ("You and I, like a tree with two roots"). Morgenstern, *Werke,* 700.

56. Steiner, *Foundation Stone*, 73.

57. "Je crois que [Morgenstern] s'était retiré dans le Götaland dans un couvent de théosophes. Je crois qu'il est devenu, qu'il est mort théosophe. Je ne connais rien de sa vie." Guillevic, *Un brin d'herbe*, 75.

book he had undoubtedly read, since he was aware that Bachelard cites him in the book and took him as a terrestrial poet. The energy of Guillevic's poem conveys what I will call "radical trying": the botanical embodiment of an underground *élan vital*, a burrowing will that implicates itself consciously in the earth, the active process of a de-alienation. In contrast to Bachelard, who viewed the root as a figure of repose, Guillevic saw it as a figure of *volonté*.[58] The root's force resembles Heidegger's gloss of the term *Wesen*, which generally means "essence" but which he nuanced to mean emergence, coming into being, or "root-unfolding."[59] This pushing through the soil at a pace imperceptible to humans is the action of pure *physis*.

The poem is essayistic in that it tries out a series of thoughts, separated by asterisks, the philosophical questions of a meditative subject. This is a poem that thinks. The empathic free verse considers the world from the roots' point of view, asking implicitly, "What is it that roots want?" and dozens of other philosophical questions throughout the verses: What is the nature of time for a plant? Do the roots "know" what is going on above ground in the branches, and vice versa? Is a root a thing or a presence? What can the root teach us about human mortality? The addressee of the poem shifts throughout, as the narrating voice alternates between first-, second-, and third-person narration; between *tu* and *vous*; between indicative and imperative. The fragmentation created by the asterisks and the spaces that divide the verses gives the poem an aphoristic quality; it resembles the haiku in its concision, clarity, and natural theme.

The epistemological thematics of Guillevic's poem insist on what is knowable and how.[60] The narrating voice says of the oak, "Through him, I know"

58. He undoubtedly would have included the root in Bachelard's other book, *La terre et les rêveries de la volonté: Essai sur l'imagination des forces*, not in his book *La terre et les rêveries du repos: Essai sur les images de l'intimité*.

59. This translation was proposed by Kenneth Maly in "Reticence," 156n14.

60. The verbs "savoir" and "comprendre" are everywhere in the poem. For example:

> Est-ce que la frondaison
> À besoin de comprendre
>
> Ce qui se passe
> dans les cuisines?
>
> (Guillevic, *Relier*, 90);
>
> "Connais-toi dans la terre
> Que tu vas distiller." (98)

("À travers lui,/Je sais").[61] The root, too, is a path to knowing; in direct contact with it (*toucher*, *palper*, *regarder*), the human comes to understand that the root is both you (*toi-même*) and the opposite of you (*l'anti-toi-même*) (93). Otherness becomes a source of creative contemplation. A living paradox, the root is both hard and soft, rigid and ductile.[62] It is depicted interchangeably as a singular entity (*la racine*) and at times as a plural multiplicity (*les racines*), slipping back and forth between the categories of individual and collective. Its form is the *ratio studiorum*, a plan of study; through various lessons made available through its form and its processes, it teaches about competition, solitude, coexistence, sex, death, and the will to live. Furthermore, it reveals the limitations of knowledge:

> Vient un moment
> Où les racines sont forcées
> De constater
>
> Que désormais
> Elles n'iront pas plus profond,
>
> Que de la terre, en somme,
> Elles ne pourront pas
> En savoir davantage.[63]

61. Guillevic, *Relier*, 89. In the Blanchet edition, this passage reads:

> A qui
> sont les racines du chêne?
>
> Elles sont à lui
> Comme il est à moi.
> A peine plus.
>
> Par elles
> Il sait la terre
> Dans sa solitude
> Dans ses remuements.
>
> Et moi, je sais,
> Par lui, avec lui,
>
> Par elles.
>
> Guillevic, *Racines,* no page number

62. Bachelard called the root "living death" ("la mort vivant"). Bachelard, *Terre*, 291.

63. Guillevic, *Relier*, 100.

Profundity, as a figure for spatial depth but also as a figure for cognitive capacity, has limits. The root—as a real, botanical form and as a figure for knowledge—can only push so deep. No root has ever penetrated the surface on the opposite side of the world, through the molten core. Guillevic pushes the exercise further; the narrator in his poem takes the root as a set of instructions for how to live:

> Façon de vivre:
> Se sentir englobé
> Par l'ensemble des racines. (106)

That is, to re-implicate the human in the rhizosphere, an ecofigure for the world. The narrator expresses his alienation from the earth through a desire to live like a root:

> Je ne demande qu'à entrer,
> À vivre comme vous,
> À me vivre racine. (106)

Throughout *Racines*, Guillevic probes the temporality of the root, what Michael Marder has recently called "plant time." The temporal vocabulary—"Un jour/Beaucoup plus tard" (92), "tes heures" (94), *siècles, l'éternité, avenir, vite*, "l'éternité des temps," "d'un moment de plus tard" (98)—hints at the fact that radical time is much slower than human time, which makes the world seem fast in comparison to it. The poet asks:

> Entre le fossile
> Et le cadavre,
>
> Pour vous,
> Quelle différence? (96)

For humans, centuries separate the fossil and the corpse. We feel a temporal proximity to the latter and a temporal distance from the former. But for the plant, which has a daily contact with both, gleaning nutrients from the cadaver and minerals from the fossil, the poet wonders if time is even "of the essence," so to speak. Death is for the root a timeless precondition for its sustenance; without it, the root cannot live.

From the beginning, Guillevic's lyric economy shows an affinity with the economy of the root. The poetic voice recognizes in it bursarship (*l'économat*) and stewardship (*l'intendance*). He also identifies in their growth process a kind of labor, a term particularly charged for Guillevic, whose Marxist sympathies are found just below the poem's surface. He calls root growth an elaboration (103), a working out of form.[64] I emphasize here the "labor" found in elaboration; growth is work:

> Racines
> Elaborant le chêne
>
> Et vous élaborant
> Monument pour vous-mêmes.[65]

General work-related words like *travailler* and *travaux forcés* are combined with job-specific work vocabulary such as *distiller* or *malaxer*, which means to knead or mix. Throughout the poem, roots are transformed into laboratories and kitchens and other spaces of making.[66] Guillevic's Marxism creeps into the poem again here:

> Comment est-ce
> Dans la terre
> La concurrence?
>
> Quand, par exemple,
> Il n'y a pas
> Assez d'eau pour toutes.
>
> Quel est le règlement
> De la répartition? (104)

64. Bachelard reminds us that "all of our ancestors were agricultural laborers [*laboureurs*]." Bachelard, *Terre,* 297. The early human history of land-labor emphasizes our status as *terriens.*

65. Guillevic, *Relier*, 103. This verse recalls Ponge's formulation in "Le Carnet du Bois de Pins": "En somme, qu'est-ce qu'une forêt?—A la fois monument et société. (Comme un arbre est à la fois un être et une statue.) Un monument vivant, une société architecturale." Ponge, *Tome premier*, 363.

66. In a passage from "Végétation" in his *Le parti pris des choses*, Ponge describes all plant life, which catches the rain as it falls, as an immense laboratory: "un immense laboratoire, hérissé d'appareils hydrauliques multiformes, tous beaucoup plus compliqués que les simples colonnes de la pluie et doués d'une originale perfection: tous à la fois cornues, filtres, siphons, alambics." Ponge, *Tome premier*, 101.

How are resources distributed throughout the vegetal multiplicity? Which laws of competition govern underground? The narrator wonders hypothetically whether the root, in its ugliness, may be a figure of revolt against the trunk and foliage for which it provides (99). Without the root's labor, after all, the tree has no chance of survival. Its noble verticality relies on the laborers, *les basses classes*, the true *agriculteurs* whose working of the land feeds the higher-ups. This pattern of the acknowledgment of workers' exertions can even be seen in the angel-winged ant, the most industrious insect, that lands on the page, in possession of the truth from the subterranean factory, and who is summarily crushed:

> On se dit
> qu'elle vient de là-bas
>
> Et on l'écrase. (105)

The crushed worker, the representative of what is low, dies under the thumb of a looming figure. The root is programmed by "a law of the earth" ("une loi de la terre") to resist orderly straightness, refusing the rectitude of human-made forms like the straight Avenue de l'Opéra (93), one of the anti-barricade features of Haussman's tactical Paris.[67] Even the ownership of private property is done away with in the verses:

> Les racines,
> Pas propriétaires
>
> De la furie
> Ni de la terre.[68]

Roots become factories that run, rather than three eight-hour shifts, one twenty-four-hour shift:

> Drôles d'usines,
> Ces racines

67. Bachelard uses the uncommon word *anfractuosity* (*l'anfractuosité*) (*Terre,* 313), a state characterized by intricate twists and turns, to describe the root's form, which puts it in the same family as the labyrinth, another image of repose in his catalog.

68. Guillevic, *Relier*, 92.

Où l'on fait les trois-huit,
Plutôt le un vingt-quatre. (93)

Their life is nothing but work. In a humorous passage, the roots receive a call from the foliage asking for more titanium from their technical department (107). This is a clear reference not only to the comical buoyancy of Morgenstern's poetry but also to his idea in "The Two Roots" of the treetops and roots communicating through some subtle channel:

The whispers where the top leaves grow
are echoed in the roots below.

Guillevic's poem wonders, using military terminology, how the plant communicates with itself:

Avez-vous quelque chose
Comme un téléphone de campagne?

À l'état-major? (108)

Countless verbs in the future tense hint at the kind of wishful futurity involved in Leftist utopianism: *apparaîtra peut-être, accablera, aura, deviendront, pourront*, and so forth. When he writes,

Dans les racines
Il y a des aventures (92),

Guillevic surely has in mind the etymology of adventure, *ad* + *venire*, or what is to come; he uses *avenir* specifically as well. Verbs of aspiration fill the verses: l'espoir (91), songer (90), réussir (91), donner envie (92), s'attendre à quelque chose (97). A figure of workers' revolt, the root's unruliness thinks forward in time to a placeless place, the utopia.

The network is a central problem in this poem and in any philosophical interrogation of rootedness.

S'improviser
Un réseau de racines

Qui relierait entre elles
Toutes les racines?[69]

This hypothetical question puts forth the idea of a fully integrated, universal root system to which all plants and trees are connected, something akin to the Matrix or the web. Michel Serres, in his well-known work on the problem of the network entitled *Hermes: Literature, Science, Philosophy* (1982), writes that the human is always caught at the junction of multiple, superimposed networks and that anyone who "fails or refuses to pass like everyone else through the crossroads of these multiple connections" is seen as an outcast.[70] He continues: "Such is the case, for example, for whoever remains frozen, hung up, in the family tree, whoever fears leaving a closed paradise between two branches of a river, or whoever wants to tear apart the network which he endures as he would a prison or slavery's iron shackles" (45). The human is necessarily a networked being, and this makes it easy to project onto the root's bifurcations an image of systemic connectedness, a logistical wonderwork that incorporates every aspect of existence (sex, birth, death, nourishment, labor) into a single living image. But the root's form is less reticular than vascular; it is less a grid than a flowchart. It doesn't criss-cross like a net but instead provides paths of flux for water, minerals, and sap, each of which has a terminus. When the narrator addresses the root with this phrase—"Vous êtes/Comme des radiographies/De quelque chose en nous"[71]—he notes the similarity between the vascular, neural, and neuronal forms in the human body seen in an X-ray and the root's ramifications. But he also hints at the hidden something, the obscure "quelque chose," that is less a physical form than a kind of ontological philosophy, a similar way of being between plant and person. Something in us is purely and secretly botanical.

Bachelard classified the root among subconscious figures of intimacy, and Guillevic takes up this claim, depicting the root as an erotic entity that penetrates, touches (*palper*), and makes love with the earth in the complicit darkness. Sexual expressions fill the poem: "amours avec la terre," "un amant jamais comblé,"[72] *pénétrer, noces, couple*, "se glisser en vous." The root's love is gluttonous. The poet proposes, as an innocent question:

69. Ibid.

70. Serres, *Hermes*, 45.

71. Guillevic, *Relier*, 103.

72. Ibid., 95.

La terre,
La racine—
Un couple? (101)

As I will show later, the notion of making love to the earth—which can be found, for example, in Michel Tournier's novel *Friday*—involves an unbirthing of the human back into the land's mother-womb. Sex is the attempt to reenter the solitude of a matricial beginning, to return to a primordial state of presocial and prepolitical life. This return is not a renaissance, but a pre-naissance. The watery, amniotic pouch in which life first stirs is not forgotten by Guillevic. Given his preoccupation with terraqueous forms, displayed in the title of his first poetry collection, *Terraqué*, the reader understands that the root is the ideal figure to pursue the thematic of earth-water symbiosis. Roots are aqueducts through which all water has passed:

Toutes les eaux
Ont passé par vous,
Racines. (101)

The poet imagines that the best way to understand the root's power of transubstantiation is to slip into it like a scuba diver:

S'il suffisait
De se glisser en vous
Comme un scaphandrier

Pour savoir comment
Vous transsubstantiez (103)

This strange image of embodiment, a self-contained figure slipping into the liquid sacrality of the root, shows the fluid penetration of one body into another.

There is other evidence of Guillevic's residual Christianity, or his radically religious atheism, as Levertov put it. He mainly infuses the plant's process of change with a mystical vocabulary, such as transubstantiation and apotheosis, which can mean "deification."[73] Processes of transformation, such as

73. The poet describes where apotheosis happens:

Passent les molécules
De la terre aux racines.

distillation or the tree's molecular making of an apple,[74] are inexplicable and miraculous. The way the poet describes the plants' eating and seduction of the earth resembles Simone Weil's language of *nourriture spirituelle*. The roots believed their work would produce something like the Big Bang, not simply a trunk and foliage; they would have preferred to make something that lasted for eternity (97). Even in its ugliness,[75] the root continues to spin out supernatural beauty in the form of flowers and leaves. The tree's miraculous verticalization of water, whose defiance of gravity is made possible by the trunk, is only one of its many marvels. The theme of repetition, of constant and steady self-proliferation, depicts the root as an eternally engaged maker of miracles, repeating the process of life-making without needing instructions on how to do so. Making life is built into the genes of the plant; it is a botanical designer, a creator of sorts. The word *humus* (112), which links the human and the soil and is a very common word in the lexicon of rootedness, makes an appearance in the poem. The plant makes life from soil, just as the God of Genesis does.

Racines is a poem of reflexivity and reciprocity. It abounds with pronominal verbs, including reflexive, reciprocal, and idiomatic pronominal verbs: *se dire, se quitter, se connaître, se retrouver, se rencontrer, se faire, se refuser, se*

C'est là que se décide
L'apothéose

D'éléments anonymes
Qui deviendront la feuille,

Que deviendront la fleur,
Le scandale des couleurs.

Ibid., 96–97

74. The poet stands in awe at the process of a molecule or metalloid making its way from the ocean to become part of the composition of the apple:

Cette molécule, voici
Qu'elle se retrouve

Dans la pomme

Qu'offre l'horizon
À la nuit qui vient.

Ibid., 99

75. Base abjection of the root is manifest in its resemblance to medieval monster or grotesque, contorted faces (ibid., 98) or to gnomes (104).

chanter, se glisser, s'enfoncer, se taire. At times, it is difficult to tell whether the actions are being performed by the object on itself or mutually between two objects. The poem also includes several tonic personal pronouns attached to *même*, either as reflexives or emphatics: *toi-même, vous-mêmes, eux-mêmes*. All of this reflexivity and reciprocity shows the extent to which Guillevic is sensitive to the correlations between the natural world and the human-made one. The unsureness of a partnership between plant and person is communicated through terms of doubt (*probablement, peut-être*) and many interrogatives; the poetic voice asks exactly twenty-five questions in the poem. The poem, then, doesn't know; it asks.

Of all the senses, sound is most emphasized in the poem. Bachelard cites Guillevic's poem "Carnac" from *Terraqué*, which reads: "Les forêts le soir font du bruit en mangeant,"[76] an image that recurs in *Racines*. Addressing the roots directly, the narrator suggests that silence is what they seek in the earth, but they create their own noise. Even in the oceans and rivers, the root has left its sonic signature:

> Dans toutes les eaux
> On vous entend. (102)

In one scene, the roots are depicted as so busy and making so much noise within their own bodies that they don't hear the approaching mole, grubworm, or spade. But roots have no ears. They have no tongues to tell us what they know. Somehow, the radical otherness of the root is understood most starkly in its lack of communicative power with the human. We can neither talk to nor hear each other. The root becomes a figure of the incommunicable. No one can really speak about it nor on its behalf. The poet is committed to the idea of living like a root except for the fact of its silence:

> Pour cela,
> Je suis prêt à tout.
>
> Sauf à me taire,
> Sauf à vous vivre sans le dire.
>
> Vivre sans dire
> Ce n'est pas vivre. (106–7)

76. Guillevic, *Terraqué*, 50.

The poem closes with an enigmatic reference to the challenge of dialogue:

On ne se quittera pas
Comme ça.

Il a fallu du temps
Pour oser se le dire. (112)

By addressing a presence that will never receive the message, Guillevic acts out the impossible ambition of real communion with the natural world. This chasm between humans and nature can be filled neither by words nor by actions. However, in contrast to Celan's radical abyss, the abyss that makes an appearance in Guillevic's poem remains a space of possibility and empathic projection:

Qu'est-ce qui se passe
Pour une racine

Quand elle débouche
Dans l'air d'un abîme? (96)

The narrator tries to imagine what it must be like as a root to encounter a cavity in the soil, which he takes as breathing room underground and a break in the dirt's seeming infinity, a moment of difference within a far-stretching sameness. These gaps, called *galeries* toward the end of the poem, are aspirations, hopeful pockets of breath where the poet allows himself to imagine he has more in common with the root than he actually does. Plant empathy was also a feature of Ponge's work; he imagined leaving the human world for the plant world and staying there forever.[77] Poetry is one of the best places where the anxieties about our divorce from nature can be given voice. Guillevic shows an unalienated entity to which we may look as a model for incorporating ourselves into the earth. The root is a figure of haptic involvement in the world.

77. Ponge writes, "Quitter ma tête, descendre au nœud de l'être, situé, enfoui dans l'ombre, sous quelques centimètres de terreau. . . . Oui, c'est cela: j'y suis; que j'y reste toujours!" Ponge, *Nouveau recueil*, 109.

THE AWKWARD HUMAN: LEVERTOV AND ECOLOGICAL ALIENATION

The American poet Denise Levertov (1923–1997) considered both Celan and Guillevic among the great poets whose "[darkness] must be within them. It's a question of weight."[78] In the late 1960s, she published her translations of some of Guillevic's poetry into English, and she also wrote a late poem called "Thinking about Paul Celan," in which she calls him

> Saint Celan,
> stretched on the cross.[79]

She published a poem in 1982 that conveys the impossibility of rerooting, particularly the sort envisioned by Simone Weil and Martin Heidegger, which is analyzed in detail in chapter 4. As a bridge figure whose root poem hovers somewhere between Celan's and Guillevic's, Levertov discerns the human's propensity to deracinate itself from the earth, which created the conditions of its very existence. Dorothy Nielsen, who describes some of Levertov's work as "antianthropocentric," claims that these poems are marked by a sensitivity to a specifically Hasidic ontology.[80] Levertov was influenced by Martin Buber's work, even borrowing his signature "I-Thou" in her poem "The Cat as Cat." In *I and Thou*, Buber stages a reciprocal contact between human and tree,

78. Levertov quoted in Hollenberg, *Poet's Revolution*, 433.

79. Levertov's Christification of Celan might have something to with her personal history with Judaism and Christianity. Her father was a Hasidic Jew who converted to Christianity and became an Anglican priest. Of this connection, she wrote,

> My father's Hasidic ancestry, his being steeped in Jewish and Christian scholarship and mysticism, his fervor and eloquence as a preacher, were factors built into my cells. . . . Among Jews a Goy, among Gentiles (secular or Christian) a Jew or at least a half-Jew (which was good or bad according to their degree of anti-Semitism); among Anglo-Saxons a Celt, in Wales a Londoner who not only did not speak Welsh but was not imbued with Welsh attitudes; among school children a strange exception whom they did not know whether to envy or mistrust: all of these anomalies predicated my later experience.
>
> Levertov, *New and Selected Essays*, 258, 260

80. Nielsen, "Prosopopoeia," 698.

making the two equals in their taking account of each other.[81] Nielsen argues that Levertov's ecological poetry evinces her "belief that nonhuman entities are presences rather than objects politically as well as poetically."[82] Despite her appreciation of Guillevic's poetry, she resisted his tendency to project upon the tree the kind of subjectivity he grants it.[83]

Levertov may have had Celan's radical abyss and Guillevic's poetry of radical trying in mind when she wrote these verses titled "Re-Rooting":

> We were trying to put the roots back,
> wild and erratic straying root-limbs,
> trying to fit them into the hole that was
> cleancut in clay, deep but not
> wide enough; or wide but too square—trying
> to get the roots back into the earth
> before they dried out and died.
> Ineptly we pulled and pushed
> striving to encompass so many rivers
> of wood and fiber in one confinement without
> snapping the arteries of sap, the force
> of life springing in them that made them
> spring away from our hands—
> we knew our own life was
> tied to that strength, that strength we knew would
> ebb away if we could not find within us
> the blessed guile to tempt
> its energy back into earth,
> into the quiet depths from which we had
> rashly torn it, and now clumsily
> struggled to thrust it back not into sinuous corridors
> fit for its subtleties, but obstinately
> into an excavation dug by machine.

81. Levertov writes of her "renewed interest in Buber through the Hasidic ideas with which I was dimly acquainted as a child." Levertov in Allen, *New American Poetry*, 441.

82. Nielsen, "Prosopopoeia," 701.

83. Nielsen, in her reading of Levertov's poem "The Life around Us," writes, "[Levertov] reminds us that the trees have no brain cells; therefore, terms such as 'knew' are anthropomorphic projections. At the same time she insists on telling what no language can tell: what happened to a nonhuman entity." Ibid., 699.

And I wake,
as if from dream, but discover
even this digging, better than nothing,
has not yet begun.[84]

Through a lexical field of natural wetness (rivers, arteries, sap, ebb), she depicts the human as a shriveling form, drying up after its own withdrawal from the source. The human, unlike the root, extricates itself from its ground. Even with tools made for digging or re-implantation, the human can only clumsily attempt to put itself back in the notch it deserted. The people, seeing themselves as godlike, are destined to fail in their imitations of creation's intricate bonds. The tension between human self-extrication and the intricacies of nature's original forms hinges on this syllable "-tric," which implies the *tricae*, or entangled perplexities or tricks and turns, simultaneously inimitable and inescapable. Restoration of an original implication, an enfoldedness in the soil, is impossible. Levertov's poem puts a global "we" at odds with itself as it looks where it was, knowing it can never go back. The impossible return plays out in the poem through verbs such as "trying" and "striving." The obstinate ineptitude with which the species attempts to reintegrate itself, trying to dig back in as artfully as the roots, shows that it can be nothing if not awkward in the world. George Steiner has used the expression "penitential ecology" to describe the guilt-induced human exertion to undo its own environmental damage.[85] This existential awkwardness seems incurable, since even the gouging out of a provisional space—"better than nothing"—is only a dream vision. The human is simply an ill fit for the planet. The heavy indentation of the poem's last stanza creates the unfillable gap between the work the narrator imagined was happening (for lack of a better solution) and the actual nonwork that keeps humans from even bothering to try. Can you imagine trying to refit an uprooted tree perfectly in its old ground, placing each root filament in its proper place? The task is impossible.

Levertov depicts the root in one of its most important iterations: a figure of mourning for a lost connectedness with the planet.[86] When she wrote, "De-

84. Levertov, *Poems*, 261–62.

85. "Today, penitential ecology and attempt at reparation, probably futile, are a mounting element in social sensibility and the politics of disgust." Steiner, *Martin Heidegger*, 136.

86. Romain Gary, in his novel *The Roots of Heaven* (*Les racines du ciel,* 1956), explores the extermination of African elephants in the mid-twentieth century and expresses a sensitivity to the distance that separates the human from the natural world:

forestation is a kind of protracted trench-warfare,"[87] she portrayed a species at war with itself and with the planet. As Gregory Bateson famously put it, "The creature that wins against its environment destroys itself."[88] The uprooting or razing of the forests is not a distinct problem from human self-eradication. If people so readily botanify themselves, thinking and living as rooted creatures, how can a tree torn from its ground not elicit some kind of irritation in the person who watches it? Why this particular limit of the imagination? Ecologically inflected philosophies support the notion of a plant-human symbiosis and thus consider the irresponsible felling of a tree nearly as serious as an assault on another human, as Levertov did. The human who uses the world's resources without following what seems to be nature's rule of taking only what is needed—I have yet to see a morbidly obese tree—goes against nature, decadently *à rebours*, and is destined to die. The root *cadere*, "to fall," is found in the word *decadent*, used today in commercials to sell luxury chocolate or other forms of excess. We are regularly invited to decline through overabundance.

A fascinating example of the kind of ecological severing that Levertov describes can be found in the early-twentieth-century botanical photos by the German photographer Karl Blossfeldt. There are few better cases of technology used to paradoxically celebrate and destroy nature, thus marking the ever larger rift between humans and the natural world. One of the most significant aspects of the root is its usual unseen-ness. At least in the Western world, the tree as a whole suffers from a general invisibility because of its vast proliferation in most populated areas; there are so many, we cease to see them.[89] But the root is more literally hidden, which means that for the human, it is in many

---

> I took a good look at those dogs, out of whom gelatine and soap would be made, and I said to myself: You wait a little, you human master race, I'll teach you, I'll teach you to respect life. I'm going to have it out with you, and with your gas chambers, your atomic bombs and your need for soap. . . . It wasn't worth while to stand up for this or that separately, men or dogs—it was essential to attack the root of the problem, the protection of nature. . . . Islam calls that "the roots of heaven," and to the Mexican Indians it is the "tree of life"—the thing that makes both of them fall on their knees and raise their eyes and beat their tormented breasts. . . . Our needs—for justice, for freedom and dignity—are roots of heaven that are deeply embedded in our hearts, but of heaven itself men know nothing but the gripping roots.
>
> Gary, *Roots*, 172–73

87. Levertov, *New and Selected Essays*, 165.

88. Bateson, *Steps to an Ecology*, 501.

89. Vilém Flusser writes, "A curious fact: trees are almost invisible. Every attempt to contemplate them proves it." Flusser, "Cedar," 35. Of this "partial invisibility" he explains that our

ways the most neglected part of the plant. When asked to draw a tree, a person is likely to leave the roots out, considering only its visible parts as integral to it.[90] The root's problem of low visibility arises again and again in the body of texts studied here; it is unsightly in more ways than one. Something about its omission from our field of vision makes it all the more enticing as a screen upon which to project fanciful ideas about it. In the early twentieth century, Blossfeldt, a central figure of the *Neue Sachlichkeit* (New Objectivity) movement in photography, attempted to make people see the plant again, but not its roots. His botanical images show a wide variety of immaculate and nearly architectural forms in the plant world, suggesting that all forms of human-made beauty drew their inspiration from these paradigmatic forms. In his *Urformen der Kunst* (literally, "*Original art forms*," but translated as *Art Forms in Nature*, 1928), black and white photographs show in detail the beautiful symmetries and contours of various species of plants, singling out an individual stem, bud, or pod and highlighting the forms that artists have borrowed for their furniture, bridges, and cathedrals. In an introductory essay to the collection, the art collector Karl Nierendorf writes that all the features of European architecture through the ages "trace their original design [*Urbild*] to the plant world."[91] But in Blossfeldt's large corpus of botanical photographs, no roots are to be found. Clearly, the root's riotous form and unclean lines were incompatible with his aesthetic vision of pure form. He was attracted instead to the parts of nature that looked *designed*. This leads to an important observation about the root's unsightliness: it does not resemble the craftsmanship of an artisanal god. One could imagine that Mother Nature herself hid the root underground because its ugliness was too overwhelming to bear. What is ugly

deep prejudices, which project religious, sexual, or other metaphorical ideas onto the tree, keep us from seeing it at all (36–37).

90. The renowned British botanist Agnes Arber wrote, "In the early days of herbalism, the underground part of the plant attracted a good deal of interest, because of its real or supposed medicinal qualities; but, since concern with this aspect has waned, the attitude of botanists to the root has become step-motherly. This is perfectly natural, since, owing to its burial in the earth, the study of the root is fraught with inconvenience; but the resulting neglect is regrettable, as it fosters a tendency to think of the plant, in relation to the environment, in terms of the above-ground parts only." Arber, *Natural Philosophy*, 131–32.

91. "The fickle delicacy of the Rococo ornament, as well as the heroic severity of a Renaissance chandelier, the mystically entangled tendrils of the Gothic flamboyant style, noble shafts of columns, cupolas and towers of exotic architecture, gilded episcopal crosiers, wrought-iron railings, precious scepters, all these shapes and forms trace their original design to the plant world." Nierendorf in Blossfeldt, *Art Forms in Nature*, 6–7.

is cropped from memory and perhaps consciousness. Nature is not uniformly beautiful, contrary to our wishes.

We might call Blossfeldt's photos "plant portraiture," since they treat the plant photographically like a human subject who has come to pose in the studio. But for these photos to happen, the plant had to be torn from its habitat, a fact denied by Blossfeldt's omission of the roots or other evidence of this tearing. The minimalism of his photographs, which show plants plucked from their context and put on display before a neutral backdrop, refuses the natural world's will to contextualize. In nature, context is everything. All natural life belongs to a finely tuned ecosystem; only the human has unsituated itself, bringing some other species along with it through domestication. Nierendorf takes a celebratory position toward new technologies and claims that they bring us closer to plants.[92] He even implies that new technologies like "mechanical buildings, motor-cars, aeroplanes, as well as the film, the radio, and photography" (6) are extensions of the same creative evolution that led people to imitate plant forms subconsciously in architecture. These technologies are simply examples of "the creative spirit manifesting itself in new forms" (6). From this perspective, modernity is not at odds with nature. On the contrary, the same *élan créateur* that impels plants to improvise their elegant forms manifests itself in all that is made by the human. In short, technological developments are extensions of the botanical creative force. The vegetal element of ourselves manifests itself in the forms we reproduce, imitative of their already mastered masterpieces.

This presents a strange paradox: if Blossfeldt's camera is the next phase in a creative evolution of which plants were the earliest step, it also destroys plants in order to do its work. Nierendorf saw in the images a step toward the realization of "a new unity in all spheres of Life, of Art and of Technics" (8), but this kind of unity preserves the plant only as a dead index, not as a living entity. The real plant was killed for art's sake. With his title *Urformen der Kunst*, Blossfeldt suggests that plants provide the *Ur*-forms of art, that for any form the artist can imagine, nature has already provided the prototype. Yet,

92. He argues, for example, that the "new type of man" who is drawn to the sunlight and the outdoors, competing in sports and driving fast in his car through the open country, "aims at active and immediate union with Nature." Ibid., 5. This idea of fast youth in nature has clear fascist overtones in retrospect. The Futurists and the National Socialists celebrated the young, heroic man, and the latter placed particular emphasis on this young man's communion with nature, for example in the scoutlike outdoorsiness of the *Hitler Jugend*. Nierendorf also celebrates slow- and rapid-motion film and close-up photography, which allow for a new kind of plant study.

by cutting up the plants and leaving the roots out of the frame, he ruptures the continuity implied between contemporary forms and their primeval paradigms. Through these photographs, several common associations of rootlessness come together. To be rootless is to lack context. It involves the anxiety of being the only species that can maintain memory from one generation to the next, that can create technologies to destroy all other species, that can dramatically alter ecosystems for its own benefit. This rootlessness can take the form of an ecological alienation, with people fearing their permanent exile from nature. It can take a social form, in which the rootless individual cannot be integrated into a polity, a community, a family, or some other form of social organization. It can also take form as an existential rootlessness; postmodernism, for example, is often imagined as the reaction to a feeling of historical placelessness, a lack of real understanding of traditions, and a break in cultural continuity. In the digital age, the preference for the impalpable over the palpable has contributed to this existential rootlessness. And contemporary irony, especially in its most hyperbolic forms, often relies on the absolute subversion or deletion of context.

Denise Levertov's "Re-Rooting" describes beautifully the ways that technology—in the form of the camera, the trench-digger, the pesticide—has decontextualized the human, giving the impression that all problems can be solved through scientifically informed mechanisms, yet doing nothing to alleviate the generalized sense of alienation. The plants in Blossfeldt's pictures are beautiful, preserved in time as indexical images, deracinated and emptied of life. Once they are extracted, it is impossible to imitate nature's original contextualizing genius. In contemporary life, as we collectively try to "put the roots back," we realize that our collective energies are spilling out and that our tools and toys are of little use to us in conserving them. Poetry might very well be the best medium for expressing uprootedness as the increasing absence of context, giving a language to what photography can only display wordlessly.

*

Four poets, four distinct roots. Ponge, in his openness to let the plant change his mind and by turning his object of study this way and that, trying to think through the plant's own experience of the world, sees clearly what distinguishes it from the human. People are nomads in comparison to the plant. Do we really want what the plants have? Would we rather have the world come to us? In our increasingly vegetative state, in which we access the remote world through a screen, we have taken on something of the plant's existence, which requires everything to come to it. But these images and sounds are only simulacra; were we to put a tree in an IMAX theater showing the film of a forest, the tree would not recognize this world as its own. The kind of embeddedness the plant requires in order to thrive cannot be replaced by images. Our artificial vegetative life might seem to want to make stillness a virtue, but it is the ultimate paradigm of decontextualization. The trees' immobility is what turns forest into context. In the case of Celan, while I believe his poem is an attempt to reconcile through poetry the gap between what should have come to pass and what actually did, it conveys much of the same longing as Levertov's ecological elegy, which tries to fill an absence with presence. Celan wants to wake up from his nightmare; Levertov wants her dream to continue. But in both cases, there is this unappeasable hole, the "o" in Celan's page or the awkwardly shaped hole—"deep but not/wide enough; or wide but too square"—in that of Levertov. Guillevic's root is content with its vocation as root, which it lives as such; pushing through the terrestrial darkness is the root's calling. It is pure will in botanical form. If there is a hole in his poem, it is the communicative breach that separates plant from person. We will never know what the plant knows. While the poet longs to be more rootlike, this is more of a thought experiment—"How would it feel to be more radical?"—than an ardent wish. In this sense, his hypothetical question resembles Ponge's essayistic treatment of the plant. To a greater extent in Guillevic's poem than in the others' works, the human and the root are mapped onto one another as entities with similar bodily and intellectual needs and wants. A certain harmony between plant and person can be found in his verses. But there is a risk to the literalization of plant-human metaphors. A marked difference exists between loving land and loving a specific plot of land. The celebration of the homeland in a regionalistic or nationalistic way can very easily tip over into a politically dangerous, proprietary miserliness, which was the case, for example, in Germany under the National Socialist regime. In many contexts, it is difficult to use the word *Heimat* without triggering images of genocide. The latent political implications in the root poems of Celan, Guillevic, and Levertov all point toward very real instances of twentieth-century brutality, exploitation, land-worship, and land-devastation. In the following chapter, I show how European culture relies on the belief that transcendence in most of its forms is predicated on a firm rootedness. A tree's verticality is owed to its underground reinforcement.

These twentieth-century poets have illustrated the wideness of the root's perimeters as a metaphor. Botanical immobility captured the imagination of Ponge; by thinking through it as a hindrance and a gift, he showed the limits to the human desire to imagine the self as a plant. Paul Celan and Eugène

Guillevic both invest the figure with potent symbolic power but in nearly diametrically opposed ways. Celan's root is dead—or, better, *void*. It is an almost total abstraction, a figure for what is virtually unsayable and the nonform of a future that will not come to pass. Guillevic is much more of a literalist, speaking of what the reader might imagine as a very real root that becomes the departure point for a series of thought experiments staging an empathic encounter between person and plant. His root is very much alive, and anthropomorphically so, becoming alternately a worker, a cook, and a lover. Denise Levertov, an avid reader of both Celan and Guillevic and familiar with Ponge's poetry, crafted a poem about the ontological awkwardness of the human in the natural world, trying to replant itself back in the environment to no avail. Rootlessness as a metaphor for ecological alienation is one of the most widespread of its uses in the twentieth century. Having cut the umbilical cord from Mother Nature and possessing the mental capacity to realize it, we find each day in the news evidence of the graceless attempt to reverse this severing. But, as in the case of the person who looks at the plant's immobility with a kind of pity, there is a secret gladness in our independence from nature. The secession of culture from nature, the great Cartesian split, has perhaps fatigued the human who senses but cannot repair this uprooting. People take themselves willingly out of context more and more.

Indeed, the four poets show that, perhaps above all else, people desire a context on which they can depend. Context means wovenness into a phenomenon, be it nature, family, or history. This is what Ponge finds appealing in the immobile plant; it is absolutely never out of context. For Celan, context means maintaining a position in a centuries-old lineage and in an ineradicable culture. For Guillevic, the plant's active self-investment in its vocation and its habitat make it the ultimately contextualized entity. And for Levertov, to be contextualized is not to look at nature from outside it like an awkward voyeur but to have never stepped outside it. When anything is taken out of context, it is inevitable that meaning will be lost. These questions continue in the next chapter, in which I introduce the idea of rooted transcendence. In the Western imagination, no one can transcend without roots.

CHAPTER 3

Roots and Transcendence

To live like a tree! What growth! What depth! What rectitude! What truth!

GASTON BACHELARD, *La terre et les rêveries du repos*

No tree, it is said, can grow to heaven unless its roots reach down to hell.

C. G. JUNG, *The Collected Works of C.G. Jung*

To walk in northern California through a grove of sequoias, the world's largest trees, is to move among wordless giants. They compel the human, facing such quiet immensity, to acknowledge her smallness. In Muir Woods in Mill Valley, California, the cross section of a Redwood felled in 1930 has been labeled with significant historical events that correspond to the tree's growth rings: the Declaration of Independence (1776), the discovery of America (1492), the signing of the Magna Carta (1215), the Battle of Hastings (1066), the birth of the tree (909). This time-keeping tree fascinated both Alfred Hitchcock, who included it in his film *Vertigo*, and Chris Marker, who cited Hitchcock's use of it in his own film *La jetée*. For over a millennium, this singular living entity had inscribed and held its space. The same can be said of few empires. Challenging gravity and the winds, the arboreal being reached upward through time and capitulated only under human duress. What had made possible the impressive stature of this tree? Many factors, of course, contributed to its survival, but its height and breadth were owed largely to the anchor and secret repository of energy below ground: the roots. The tree's verticality depends on the tenacity of its root structure, which spreads and clutches the soil, siphoning water and minerals from it. *Rootedness is the precondition for verticality, or, in metaphysical terms, transcendence.*

While the meaning of transcendence varies from discipline to discipline, its etymology contains within it the basic notion of surpassing or overcoming something. Like the word *surmount*, distance and height are implied in its etymological components *trans-*, which means "beyond," and *scandere*, which means "to climb." In theological terms, transcendence describes the nature of an experience or even an existence—that of God, the angels, and the dead who have ascended to the heavens—beyond the reality of the material world. To transcend is to overcome materiality, to pass ontologically from a physical state to a metaphysical one. Immanence, the opposite of transcendence, describes that which belongs to the material world as well as manifestations of God on earth, perceptible to humans. The miracle, for example, is immanent evidence of God's existence and agency that makes itself available to the human witness. To transcend is to go beyond, upward, and toward the yonder along a vertical axis, away from that which is materially given. It is a move toward the qualitatively superior, a liberation from the flaws, shortcomings, weaknesses, and evils associated with human fallibility. The varieties of transcendence, then, include the most familiar spiritual type as well as an intellectual overcoming of the cognitive limits of the human and a historical or anthropological type that allows collectives to triumph over their seemingly innate discordant and bellicose nature. Each of these types of transcendence involves a fundamental improvement of the human, either as an individual or as a collective, to the point of overcoming the human altogether, undoing the ontological and epistemological limits imposed by the burden of embodiment. The possibility of transcendence is predicated on the successful establishment of a stable base from which to move upward along a vertical axis. Metaphysical ascension requires an immovable and nourishing ground from which to begin. Among writers and thinkers in France and Germany of the twentieth century, the metaphor of choice to describe this process—it is indeed a process, as it requires applied effort and time in which to unfold—is the notion of rootedness, which invites the implicit botanification of the human. To imagine humans as rooted, one must imagine the person as a plant.

One of the most familiar iterations of people imagined as plants can be found in Ovid's *Metamorphoses*, but Greek precedents exist as well. In his *Timaeus*, Plato describes the human as a celestial plant whose roots are fixed in the firmament. The human's vertical bearing is due to this rootedness: while simultaneously sprouting down out of heaven and striving toward it with the uppermost, soulful part of the body, the human is a bridge between the earthly and the divine. For Plato, our head, with its intellect and soul, is metaphorized as a root. He writes: "We should think of the most authoritative part of our soul as a guardian spirit given to each of us by god, living in the summit of the body, which can properly be said to lift us from the earth towards our home in heaven, as if we were a heavenly and not earthbound plant. For where the soul first grew into being, from there our divine part attaches us by the head to heaven, like a plant by its roots, and keeps our body upright."[1] Plato's version of gravitational pull makes heaven, not earth, gravity's center. The godly suspension of the human is contingent upon the embeddedness of the soul in the extraterrestrial plane, as the "up there" draws people toward it. Our uprightness, figuratively and literally, is attributed to the divine power's planting of the human in the celestial realm. This early metaphor, I argue, is lodged deeply in the collective subconscious of Western civilization and accounts for many stubborn notions of provenance, belonging, and the relationship between the material and the spiritual. Formulated in antiquity, the relationship between root, verticality, and God continues in the Christian tradition and has particular resonance in the French and German philosophical, literary, and political realms. This triangulation is unmistakable, for example, in the notion of Jews as a rootless, and thus unrighteous, people, reflected in the long history of anti-Semitism in Europe. Paul Claudel, the father of neo-Catholicism in postwar France, in his *A Poet before the Cross* (*Un poète regarde la croix*, 1935), suggests that Christ's redemption of humanity was rooted in the Garden of Eden; in his depiction, the Tree of Knowledge transforms into the crucifix, thus making a connection between the tree, the origins of humanity, and transcendent salvation. Simone Weil argued in the 1940s that the easy capitulation of France to Germany in World War II was caused by a collective loss of roots; the path to transcendence, in her view, is the cultivation of new ones. Finally, shifting from the realm of theological or political discourse to that of philosophy, Heidegger's notion of *Bodenständigkeit*, or groundedness, confirms again the link between stability down below and transcendence up above in Western thought. His ideas on groundedness influenced French poststructuralist thinkers in the second half of the twentieth century.

The first part of this chapter discloses many more examples of the link between root, verticality, and transcendence, emphasizing the ubiquity of this tripartite symbolic configuration in the French and German cultural *patrimoine*. The second part of the chapter treats the very specific Platonic notion of the "invertibility" of the plant as a figure for metaphysical transformation, which persisted in France throughout the centuries. For the Encyclopédistes, the botanist Duhamel's successful eighteenth-century experiments in uprooting and replanting willow trees' branches into the soil upside-down become a figure of the seemingly peculiar but absolutely necessary task of the philosopher. The soul of Paul Valéry's Monsieur Teste—a fictional character who is

1. Plato, *Timaeus*, 88.

an embodied thought and, like Valéry's Lucretius, "a Plant that thinks"[2]—is described by the character's wife as "some peculiar plant whose root, and not foliage, would thrust against nature, toward the light";[3] and, in Michel Tournier's retelling of the Crusoe story, Friday anticipates and "models" Robinson's metaphysical transformation by uprooting and inverting some of the island's plants and by disguising his own body as a tree. While such botanical inversions may be read as mere examples of the improvisational potential of the plant and the human to adapt in the face of predicament, they can be read more compellingly as metaphors for metaphysical conversion, as I will show.

VERTICALITY AND THE ROOT

What is given vertically incites awe.

ANTHONY STEINBOCK, *Phenomenology and Mysticism*

When Anthony Steinbock states, "Verticality is the vector of mystery and reverence; horizontality is what is in principle within reach, graspable, controllable,"[4] he points toward a tradition in Western thought that associates upright orientation with the sacred and lateral orientation with the banal and the worldly. God is not laterally accessible; the locus of the holy is over our heads. As Pierre-Henri Hadot notes, the Greek tradition that connects contemplation with looking upward toward the stars participates in the pattern of verticality associated with the relationship of the human to the divine.[5]

2. Valéry, *Œuvres*, 192.

3. Valéry, *Monsieur Teste*, 26.

4. Steinbock, Phenomenology, 13.

5. Hadot, "L'homme, 'plante céleste,'" 81. He cites specifically Ovid's *Metamorphoses* and Philo of Alexandria's *De plantatione*. From Ovid:

A creature of a more exalted kind
Was wanting yet, and then was man design'd;
Conscious of thought, of more capacious breast,
For empire form'd, and fit to rule the rest:
Whether with particles of heavenly fire
The God of nature did his soul inspire,
Or earth, but new divided from the sky,
And pliant, still retain'd the ethereal energy,
Which wise Prometheus temper'd into paste,
And, mix'd with living streams, the godlike image cast.
Thus, while the mute creation downward bend

Perhaps the Greeks' conception of the soul as breath (pneuma [πνεῦμα]) accounts in part for the lightness, invisibility, and upwardness found in Western notions of spirituality. Phenomenologist Erwin Straus attributes this pattern to the human's exceptional upright posture and to the lengthy period and strenuous exertion necessary for a human child to begin to move about in the world vertically. Unlike breath or the heartbeat, which function automatically, sitting, standing, and walking upright are challenges to be met. The physical burden of gravity is thus transposed into a metaphysical challenge in the human subconscious.[6] In Straus's convincing account, verticality as transcendence is deeply

---

Their sight, and to their earthly mother tend,
Man looks aloft, and with erected eyes
Beholds his own hereditary skies.
From such rude principles our form began,
And earth was metamorphosed into man.

Ovid, *Metamorphoses,* 6

From Philo's *De plantatione* (*About the Planting of Noah*):

And, indeed, God has caused plants to grow with their heads downwards, having fixed their heads in the deepest part of the earth; and having drawn up the heads of the irrational animals from the earth, he has set them up high on long necks, putting their fore feet under their necks as a kind of foundation. But man has received a pre-eminently superior formation. For of all other animals God has bent the eyes downwards, so that they look upon the ground; but on the other hand, he has raised the eyes of man so that he may behold the heaven, being not a terrestrial but a celestial plant as the old proverb is.
Philo, *Works,* 419

This proverb, included as a footnote, is Ovid's, cited above:

Pronaque dum spectant animalia cætera terram.
Os homini sublime dedit: cœlumque tueri.
Jussit et erectos ad sidera toller vultus. (419)

See also Claudia Baracchi, "Looking at the Sky: On Nature and Contemplation," *Research in Phenomenology* 39 (2009): 13–28. It is interesting to note that the etymological root of the word *consider* can possibly be traced to the Latin *sidus*, or constellation.

6. The direction upward, against gravity, inscribes into space world-regions to which we attach values, such as those expressed by high and low, rise and decline, climbing and falling, superior and inferior, elevated and downcast, looking up to and despising. On Olympus, high, remote, inaccessible, and exalted, dwell the Homeric gods. On Mount Sinai, Moses receives the Ten Commandments. Below, in the depths, is Hades and the world of shadows. There, also, is the Inferno. However, such evaluations are not unequivocal. "Base" (adjective) and "base" (noun) have, in spite of their phonetic resemblance, different etymological roots and opposite

embedded in language and is experientially grounded. He writes, "When we lower our heads or kneel in prayer, when we bow or bend our knees in greeting, the deviation from the vertical reveals the relation to it" (146). That is, by showing reverence and lowering ourselves before the divine, we are signaling a self-relegation and yielding to the greater power. The degree of our lowness stands in proportionate contradistinction to the degree of respect toward the vertical sacred.

Christian Froidefond notes that "the notion of verticality is often associated with the notion of stability"[7] and, perhaps more significantly, that "verticality implies the action of a *force*, whether the growth of plants, the erection of a column, [or] the vertical bearing of certain animals" (257–58). In these examples, as in Straus's view, upwardness requires a challenge to gravity, leading one to make the more generalized argument that any physical resistance, including that of gravity, may be depicted figuratively as a spiritually meaningful challenge or burden. To meet challenges is to engage in the assertion of self against an opposing force. The tree's defiance of gravity by means of its stabilizing root structure can easily be metaphorized as a figure for the human's upward struggle, against gravity and through the elements, toward transcendence. These claims are consonant with the examples previously given here of why the vertical axis is of such importance in considerations of the divine. This pattern, that what is up is simultaneously unknown and superior to us, is confirmed over the centuries, from early philosophical texts to twentieth-century phenomenological studies and conceptual metaphor studies.

Straus makes a further claim about the ways in which verticality nuances the conception of human as autonomous individual and how this nuance is ultimately detrimental to the idea of community. In a community, verticality as a posture means a collection of parallel lines that never touch.[8] If we follow the analogy of human as tree and community as forest, there remains an irreconcil-

---

meanings. "Base," the adjective, is derived from the Latin root *bassus* with the connotation "short" and, later, "low"; "base," the noun, originated in the Greek root *baino*—"walking" or "stepping." The earth that pulls us downward is also the ground that carries us and gives us support. The weighty man signifies, by his dignified gait, that he carries a heavy burden but sustains it well. Upright posture as counteraction cannot lack the forces against which it strives.

Straus, *Phenomenological Psychology*, 142

7. Froidefond, "Linéarité," 258.

8. "In upright posture, we find ourselves 'face to face' with others, distant, aloof. . . . Parallel verticals do not meet. Therefore, the strict upright posture expresses austerity, inaccessibility, decisiveness, domination, majesty, mercilessness, or unapproachable remoteness, as in catatonic

able separation between people despite their clustering together in collectives, in Straus's view. While one is able to dwell among others, a real contact with other humans is impossible without bowing oneself or bending toward one's neighbor. The verticals in a cluster remain alone together. Francis Ponge also noted this vertical isolation of pine trees and personified them in his "Le carnet du bois de pins," wondering whether the trees, despite their separation above ground, maintained the same autonomy in their roots.[9]

Georges Bataille, in his essay "The 'Old Mole' and the Prefix *Sur* in the Words *Surhomme* [Superman] and *Surrealist*," a sustained reflection on the high and the low, claims that the human resembles a plant in its perpendicularity to the ground and that the human's uprightness has symbolically moral implications. The plant thrusts itself willingly toward the abject soil, which is equally tempting to the human and which stands in opposition to "spiritual elevation." In the introduction of his well-known essay "The Big Toe," he claims that the big toe separates the human from apes and that the human foot provides the necessary conditions for our erect posture. He insists on the ambivalent relationship people maintain with the ground, which represents all that is low and filthy, the place whence we came but which provides the necessary ground for our vertical transcendence. Our resistance to the ground and our antigravitational struggle away from it is ultimately undone by death, when we sink back into the earth and become it, through the process of decomposition.

In French, the translation of the expression "upright" is *droit*, which is also the word for "law" (*le droit civil*) and "right" (*les droits de l'homme et du citoyen*). As Christian Froidefond has shown, the association between rectilinearity and moral rightness was already firmly in place in Plato's time. The root *orthos*, which exists in contemporary terms such as "orthodox" ("right opinion"), is difficult to render in Latin, French, or English. The corresponding approximations in each language (*verus, vrai, true*) fall short of the original meaning, which included simultaneously the notions of the rectilinear and the vertical.[10] Froidefond draws a fascinating analogy between the words *orthos* and *plagios*, equating the first with verticality and the second with horizontality. While what is orthodox is associated with uprightness and rectitude, the

symmetry. Inclination first brings us closer to another. Inclination, just like leaning, means literally 'bending out' from the austere vertical." Straus, *Phenomenological Psychology*, 145.

9. "In the air, pine branches respect one another mutually, remain isolated, do not viciously intermingle (furthermore, this is what is quite curious, remarkable), but does the same go for the roots in the earth?" Ponge, *Tome premier*, 365–66.

10. Froidefond, "Linéarité," 279.

root *plagios* finds its way into the word "plagiarism," literary thievery, with *plagios* transforming from its connotation of flatness and spread-out-ness to the snare that is laid out to catch an animal. The plagiarist thus sets out a trap and snatches up the work of another, or catches readers themselves in his or her deception. The horizontality of *plagios* comes to represent underhandedness and deceit. Many have noted that in the left-right dichotomy, right is good and left is bad. This is most apparent in French, with *gauche* meaning left but also awkward or lacking grace of speech or movement. In Italian, right is *destra* and left is *sinistra*, which exists also in languages such as English (*sinister*) and French (*sinistre*), meaning wicked or ominous. Thus, that which is not upright or on its way to being so is necessarily encoded in the Western imagination as evil or wrong.

These abundant examples illustrate the extent to which metaphors of transcendence are experientially based and latent in language. A debate whether transcendence is an anthropologically or theologically determined human universal was triggered by René Girard's controversial *Deceit, Desire, and the Novel* (*Mensonge romantique et vérité romanesque*, 1961). Girard puts the novel form and Christian transcendence in dialogue, arguing that the novel, which began as a solidly European tradition, relies on specific elements of Christian symbolism. He claims that "Christian symbolism is universal for it alone is able to give form to the experience of the novel."[11] This symbolism primarily takes form as a pattern of transcendence inherent to the novelistic narrative. He goes on to explain, "Repudiation of a human mediator and renunciation of deviated transcendency inevitably call for symbols of vertical transcendency whether the author is Christian or not" (312). Many have disputed Girard's contentious claim; for example, Andrew McKenna argues that the desire for transcendence is an experientially informed, anthropological fact,[12] which aligns with Straus's claims that our bodily condition as humans predated and informed our subsequent relationship with the (imagined) divine. One of the novelist's skills consists in creating the illusion of telling with uncanny accuracy what we ourselves

11. Girard, *Deceit*, 310.

12. "We find this pattern of fall and redemption in Dickens and others not because it is prescribed by Scripture but because it is observed by the novelist, often in his own personal experience. This is the case regardless of whether, with Mme. de Lafayette and Dostoevsky, the writer is a believing Christian, or whether with Proust, Stendhal, and Camus, for instance, he or she is an agnostic or atheist. This is not theological but anthropological revelation, born of experience we can all understand and that the novelist can, by myriad narrative strategies, make us acknowledge as our own." McKenna, "Rorty," 306.

have experienced. What remains unclear is whether this consonance between verticality and the novelistic narrative is a specifically Western phenomenon or if this pattern can be found in similar types of narratives in non-Western civilizations. It is nearly impossible to determine a causal relationship between the two, but the fact remains that transcendent verticality is a pattern simultaneously present in Christian narrative (the Bible, hagiographies, etc.) and in the novel form. In other words, Christianity and the novel were originally both vertically, not horizontally, oriented.

But how does one ascend, either spiritually or intellectually, as a collective or as an individual? What are the prerequisites for an upward-moving life or afterlife? I argue that in the Western imagination, the often overlooked precondition for transcendence is rootedness or groundedness. In other words, verticality depends on a measure of horizontality, weight, and tenacity at holding one's ground. Such rootedness or groundedness takes many forms but is the consistent precondition for the varieties of transcendence described above.

The tree, whose verticality relies on its subterranean moorings, has remained a symbolically charged figure in Europe throughout the centuries. As Bachelard wrote, "The tree is a stabilizer, a model of uprightness and firmness."[13] Imagined as the bearer of the fruit that occasioned the fall of man, the Tree of Knowledge connects the arboreal and the epistemological in a single concept, a connection that has persisted throughout the centuries. In the seventeenth century, Descartes created a similar schema called the Tree of Philosophy, whose roots were metaphysics, whose trunk was physics, and whose branches were the other sciences, including medicine, mechanics, and morals.[14] In genealogy and the natural sciences, the tree became an essential schematic structure for depicting the relationships between people or other organic life-forms, thus organizing knowledge in a visually comprehensible way.[15]

13. Bachelard, *Terre*, 318.

14. "Ainsi toute la philosophie est comme un arbre dont les racines sont la métaphysique, le tronc est la physique, et les branches qui sortent de ce tronc sont toutes les autres sciences, qui se réduisent à trois principales, à savoir la médecine, la mécanique, et la morale; j'entends la plus haute et la plus parfaite morale, qui, présupposant une entière connaissance des autres sciences, est le dernier degré de la sagesse." Descartes, *Principes*, 21.

15. For example, Charles Darwin wrote of the tree as a helpful visual metaphor for the depiction of relationships between things and families of things *in time* in the natural sciences:

> The affinities of all the beings of the same class have sometimes been represented by a great tree. I believe this simile largely speaks the truth. The green and budding twigs may represent existing species; and those produced during each former year may

CLAUDEL'S ROOTED CRUCIFIX

The tree and its associations with life and knowledge provide a unifying thread between the Old Testament and the New Testament. This thread and, more specifically, the notion of the root, fascinated the French writer and devout Catholic Paul Claudel (1868–1955), who used biblical imagery and various biblical tableaux (the scene of the Garden of Eden, the Nativity, the Crucifixion) as hermeneutically decipherable images for understanding the Christian mystery. His vision is significant in that he provided one of the most resounding poetic voices of French Catholicism in the literary and theological history of the nation. He was elected as an *immortel* to the Académie française in 1946 just after World War II, thus validating him as the literary father of the neo-Catholic movement in postwar France. Claudel's religious fervency was the kind that takes everything as a sign of God's work and as an invitation to multiply symbols and intertwine them. I would not argue that he is characteristically Catholic or characteristically French in his particular fashioning of the root metaphor; rather, I show that his hyperbolic exploration of it shows how far a devotional metaphor can be pushed. More an exception than a rule, Claudel attempts to test the most extreme possibilities of using metaphor as a tool of conversion. His compulsive generation of images borders on the baroque,

---

represent the long succession of extinct species. At each period of growth all the growing twigs have tried to branch out on all sides, and to overtop and kill the surrounding twigs and branches, in the same manner as species in the great battle for life. The limbs divided into great branches, and these into lesser and lesser branches, were themselves once, when the tree was small, budding twigs; and this connection of the former and present buds by ramifying branches may well represent the classification of all extinct and living species in groups subordinate to groups. Of the many twigs which flourished when the tree was a mere bush, only two or three, now grown into great branches, yet survive and bear all the other branches; so with the species which lived during long-past geological periods, very few now have living and modified descendants. . . . As buds give rise by growth to fresh buds, and these, if vigorous, branch out and overtop on all sides many a feebler branch, so by generation I believe it has been with the great Tree of Life, which fills with its dead and broken branches the crust of the earth, and covers the surface with its ever branching and beautiful ramifications.

Darwin, *Origin of the Species,* 162–63

For more on trees used to organize visual information, see Manuel Lima, *The Book of Trees: Visualizing Branches of Knowledge (*New York: Princeton Architectural Press, 2014); Astrit Schmidt-Burkhardt, *Stammbäume der Kunst: Zur Genealogie der Avantgarde (*Berlin: Akademie Verlag, 2005); and Pacotte, *Le réseau arborescent.*

imitative of the relentless proliferation of some plant species. It may seem an odd comparison, but the multiplication of synonymous images—synonymous in that they communicate the same thesis—resembles the proliferation of images in Deleuze and Guattari's *Anti-Oedipus* and *A Thousand Plateaus*. These three writers allow one metaphor to bleed into the next in their maximalist prose and let the pictures do the talking. As I show in chapter 7, the authors also shared a mystical vision of the plant.

In Claudel's entire exegetical corpus, but particularly in *A Poet before the Cross*, he appropriates specific biblical events and their symbolic content as diachronic scenes in which the past, present, and future are simultaneously accessible. He takes as his writerly objective the reading of what I will call cryptographic tableaux. In these passages, word, image, and symbol combine to create a singular, decodable unit, all elements of which merge into one readable sign, a meaningful whole that short-circuits time and space. The move Claudel attempts to make in his biblical exegesis is comparable to the etymological obsessions by thinkers such as Martin Heidegger, fueled by the belief that if the origins—of signs, of words, of symbols—could be traced, the mystery, of language and of life, could be deciphered.

Much of Claudel's interpretive performance has to do with negotiating the tensions between the low and the high, in other words the poles of the vertical scale essential to the argument of this chapter. Claudel was particularly sensitive to the root metaphor and its connection to transcendence. In the work of many writers, the root is a superficial metaphor for talking about origins. However, in Claudel's work, the root is not only a necessary component of his most significant cryptographic tableaux; it is a key figure for understanding one's relationship to God, Christ, and all that is holy. For him, rootedness of humanity in original sin is both literal and figurative: the roots of the Tree of Knowledge in the Garden of Eden, a real tree that actually existed, contained within them the fall and redemption of humanity.

Claudel follows a *topos* that developed during the medieval period, in which the True Cross on which Jesus was crucified was made from the Tree of Life in the Garden of Eden.[16] In his *A Poet before the Cross*, Claudel makes a series

16. Simone Weil, Claudel's contemporary, also associates the Edenic tree with Christ's cross. Some examples of her association of the Edenic tree with the cross include these: "The tree of the cross. If we cut down the tree of good and evil, if we renounce making a choice, we then have the cross, the full weight of pure necessity. The illusory power of choice lies at the root of the compensatory imagination which prevents us from being under the full weight of necessity" (Weil, *Notebooks*, 2:380); "The rooting up of vegetative energy itself and its orientation toward

of cognitive leaps between various symbolically charged Christian images. He writes:

> What else is the serpent save a simplified diagram, the breath in a sheath, the root of the soul in the flesh, the humiliated spirit grovelling on the ground, the spirit which incessantly schemes against the heel of our mortality and keeps us from subsisting? What form was more suitable to the Word in His final abasement when for us he bore sin and mire. *But I, poor worm, have no manhood left* (Ps. 21:7). It pleased God to pick that up and give him a *name which is greater than any other name* (Phili. 2:9), so that we should see interwoven on the cross in the same calligraphy the root of our fault and the root of our salvation. This is the Tree which for our wonder produced the *Kingdom of God*, the *seed* which the Gospel speaks of, so small that it is the smallest of all and cannot be perceived by our senses. It has now become this tremendous tree.[17]

Three concepts here emphasize the centrality of the Word to Christian dogma, each with a specific role: *Verbe* (the Word), *Evangile* (the Gospel) and *calligramme*, rendered by the translator here as "calligraphy." *Le Verbe*, or God's word, which creates as it is uttered, binds creation and speech. It is not yet fixed in time by book or writing. The Word does not tell things; it makes them. The *Evangile*, the Gospel or Holy Book, is a record of events written by man.

---

the higher constitutes redemptive suffering. It ransoms creation while unmaking it" (1:255); "The Tree of the World—the eternal fig-tree whose roots have to be severed with the axe of detachment (Gîtâ). This is the vegetative energy. The cross is of wood, but it is made from a tree that has been cut down. Adam ate of the fruit of the tree (Two birds: one of them eats the fruit . . .). The tree must be cut down and one's own dead body must be its fruit. The vegetative energy has to be uprooted" (1:298); "One must uproot oneself; cut down the tree and make of it a cross, and then carry it always; uproot oneself from the social and vegetative angles; have no native land on this earth that one may call one's own. To do all that to other people, from the outside, is an ersatz form of de-creation; it is producing unreality. But in uprooting oneself one seeks a greater reality" (1:298). Saint Bonaventura's thirteenth-century meditation *Lignum Vitae* imagines the life and death of Christ through a living tree: "Picture in your mind a tree whose roots are watered with an ever-flowing fountain that becomes a great and living river with four channels to water the garden. . . . From the trunk of this tree, imagine that there are growing twelve branches that are adorned with leaves, flowers and fruit. Imagine that the leaves are a most effective medicine to prevent and cure every kind of sickness, because the word of the cross is the power of God for salvation to everyone who believes." St. Bonaventura quoted in Carey, *Tree*, 37–38.

17. Claudel, *Poet*, 223–24.

This written word does not create; it is merely mimetic. It reproduces in written language what were supposedly the foundational moments of the Christian faith. The tellers are men, not God. The *calligramme* transforms the various tableaux of the written account into deeply charged, readable units, or sign clusters that accumulate and mean more than their discrete parts. Claudel would have been familiar with the notion of *calligramme*, typically composed of text whose distribution on the page contributes to the meaning of the words themselves, which had been explored by Apollinaire in his 1918 publication *Calligrammes*. Apollinaire, who coined the term, described *calligrammes* as "an idealisation of free verse poetry and typographical precision,"[18] as they are often arranged in representational forms (the shape of a face, the form of falling rain). However, Claudel transforms the calligram into a configuration in which the words eventually fall away, leaving only clusters of symbolic signs that merge to make one relevant unit. Claudel qualifies the cross and its configuration as a calligram that shows humanity both "the root of our fault and the root of our salvation." Here, the Tree of Knowledge rather than the Tree of Life provides the wood from which Christ's crucifix is built. This remarkable image, which mimics the earlier medieval depiction described above, contributes several new elements to the *topos*.

First, the notion of snake and root are conflated in Claudel's passage, a fusion that later interested Bachelard in his *La terre et les rêveries du repos*.[19] Claudel's description of the serpent as a simplified diagram, pattern, or blueprint ("une espèce de schéma simplifié")[20] is a figure for the *calligramme*-like tableaux that populate his oeuvre. The *schéma* involves a single decipherable unit that summarizes or encapsulates the entirety of a given message. Like fractals, a geometric pattern whose parts have the same statistical pattern as the whole, the single snake encapsulates in its form the entire story of Christ's abasement and renewal. The snake and the root share an abject nature, but in Claudel's telling, they are both mere beginnings of a story of exaltation. Real roots are nefarious to the human in that they have the power to bore through

18. "Quant aux *Calligrammes*, ils sont une idéalisation de la poésie vers-libriste et une précision typographique à l'époque où la typographie termine brillamment sa carrière, à l'aurore des moyens nouveaux de reproduction que sont le cinéma et le phonographe." Apollinaire quoted by Michel Butor in the preface of Apollinaire, *Calligrammes*, 7.

19. "[The snake] is the most *terrestrial* of animals. It is truly an animalized root and, in the order of images, it is the hyphen between the plant world and the animal world." Bachelard, *Terre*, 262–63.

20. Claudel, *Poète*, 250.

man-made structures such as walls and foundations. The twisting fibers of the root drill their way through cement and stone simply to reach a water source. The Romantic obsession with ruins points toward this capacity of the botanical to overrun and erase the architectural remnants of civilization. The root here is indeed a figure for the fall of man, albeit rectifiable. But the snake and the root are images not only for endings but for beginnings as well. The snake engenders man's life as a mortal being. Given that the word *sémence* appears shortly after the snake imagery and that Claudel's exegesis moves vertically from down to up, the phallus is clearly connoted, with the tree beginning as "the smallest seed" and swelling eventually to immense proportions. *Sémence*, the stuff of biologic origin, is also a metaphor for beginnings. Another hint at the connection between the snake and the male sex organ is the reference to Psalms 21:7: "*But I, poor worm, have no manhood left.*" The worm, the diminutive of the snake, depicts the lesser man whose member is shriveled and impotent. However, this sexualization of the snake as a base figure for the lowliness of the body is only half of the story; the other half is one of ascension and triumph over the body. Of the serpent, Claudel writes, "Thus everything, even the Serpent himself, that living insidious root entwined on all cordages and all the vessels of our soul and our anatomy, the lowest, eating dust for food, is exalted as on the end of a pike, within the divine plan" (143). In Claudel's vision, the low and the high are intimately bound; the snake and his lowly origins are transfigured and elevated, following a recurrent pattern he identifies in the Gospel, visible in Christ's humiliation and subsequent glory.[21]

Claudel also revives an important connection between roots and stars, inspired directly from Revelation 22:16: "I, Jesus, have sent my angel to give you this testimony for the churches. I am the Root and the Offspring of David, and the bright Morning Star."[22] Claudel makes a point of emphasizing this portion of the quote in the footnote: "*Ego sum radix*" (299n1). He goes on to explain: "The body was formerly here below what transforms the spirit into a face, but now the spirit is that which transforms the body through the face and above the face into a light, and through this effect, enveloped with rays and directions, it becomes breathable and intelligible to everything that surrounds it. Such is the meaning of the passage in the Apocalypse in which God promises to give the faithful star

21. "Jesus Christ tells us that the measure of His exaltation is the measure of his abasement." Bachelard, *Terre,* 224.

22. New International Version. "Moi Jésus, j'ai envoyé mon Ange pour vous confirmer ces choses dans les Eglises. Je suis la racine et la postérité de David; l'étoile brillante du matin." Martin Bible, 1744.

of the morning. For He Himself is in her the *root, the splendid morning star* from which emanates in all directions a life-giving ray" (162). The highness of the star that signaled Jesus's birth is mirrored by the low root of provenance. The nativity scene provides a tableau that juxtaposes lowness (in the form of the humble conditions of birth, among animals and in a manger) and highness (in the form of the star and the three highnesses, the Magi, who come to pay homage to the newborn). Claudel sees in the cross a configuration in which all opposites converge and resolve themselves. Using the cardinal points to illustrate the reconciliation between opposites, Claudel gestures toward the etymological root of the word "cardinal," a word used for geographic direction and for the Church's dignitaries. The root of the word "cardinal" is "cardo," meaning "door hinge." Thus that which is cardinal is that upon which everything else hinges: "The cross is God at work. It is not only His instrument, it is His active form, His extracting unifying operation, His extension between the four cardinal points: the north or zenith which is the root in the *firmament*; the south or nadir which is matter warmed by grace on which force is exerted; the arms to the right and the left are the instruments of His temporal energy" (228). Claudel recalls here the Platonic notion of the root in the firmament. The north, the direction most associated with guidance, is here not merely that which lies in the direction of the North Pole but rather that which points toward the heavens, which lead the human in times of disorientation. The south here is not a cartographic south; rather, it is downward into the bowels of the earth, where the weight of gravity and earthly burdens press toward the ground. The right and left arms of Christ on the cross are "instruments of His temporal energy." The reference to time, a man-made notion or at least a phenomenon made measurable by man, could here refer to the horizontal, ordinal continuum of history, which Christ enfolds wholly and without reserve, reaching to the left and to the right along the human plane as an embrace of humanity through time. He exists beyond the life span of his physical body to express his love toward humankind from its inception and beyond his own bodily death. Christ's sign of love is an openness to and embracing of the whole of the human race. As Claudel puts it, "Everything on the cross is consummated, not in the constitution of an inert figure, but in desire, in cooperation and in a uniting organic activity. What was broken is restored" (142). Again, the rooted cross is the singular figure in which the Fall and Redemption crystallize.

Finally, and perhaps most importantly, Claudel reinforces the connection between Adam and Jesus with the unifying theme of wood:

> Concerning wood, under which we see the Son of Man appear tottering and crushed—and which three times got the better of Him—so many writers, more

learned than I am, have written, that there is little left for me to say. Adam ate the tree with the apple and now he is one with this robe of wood which associates him to the root of each of its fibers, physical and spiritual, to the knot of its joints, to the travail of birth and death agony, to the accomplishment of its vegetable purpose. The wood, this rough cross which had not yet achieved perfect form, is everything to which God has under the circumstances entrusted the task of making man again in His image, everything which in this ambiency is born to exercise on him an action, a formative constraint. It is the framework and the director. It is the instrument of the painful harmonization in ourselves of the inner with the outer.[23]

This passage emphasizes the corporeal rootedness of the Fall and the Redemption, not only in Adam but implicitly in every living human, as offspring of the first man. By stressing that the root is a physical and spiritual one, Claudel attempts to make his argument both literally and figuratively, or rather, both physically and metaphysically. The wood draws Adam in to each of its fibers and to the knots of its joints, biologic components that belong to both human and plant, thus reinforcing the image of man as botanical being. Another obvious connection implied in this excursus on wood is the vocation of Jesus's earthly father, Joseph, that of carpentry. In the Christian imagination, Jesus was witness as a child to the constructing and artisanal crafting of wooden structures. His life expired on a structure built of the same material. Claudel chooses the word "armature," translated here as "framework," to describe the wood's function as the infrastructure for the Fall and Redemption of humanity. Claudel's unusual expression "devoir végétatif," insufficiently rendered in English as "vegetable purpose," invests the tree itself with agency and motive, thus personifying it. Perhaps Claudel attempts here to Christianize Aristotle's notion of the vegetative soul, that intrinsic force in the plant that is responsible for its growth and reproduction.[24] In his recasting of the term, he invests the tree with the ability not only to preserve and proliferate its own life, but to proliferate Life, *all Life*, through Christ. The tree's duty, from its rootedness in the Garden of Eden to its apotheosis as the cross, is one with its very material-

23. Claudel, *Poet*, 47.

24. The notion of the vegetative soul appears in Aristotle's *De anima* and is taken up by Thomas Aquinas in his *Summa theologica*. Claudel defines the difference between the vegetative and the animal in this way: "The plant is only an image; the animal is an intention" ("La plante n'est qu'une image, l'animal est une intention"). Claudel, *L'art poétique,* 156.

ity, and its message insinuates itself into the *dramatis personae* of the Christian chronicle. The tree provides the material out of which the Christian mystery is built.

In an odd footnote in this section, Claudel writes that he read the following passage "somewhere" ("lu quelque part"), without specifying the source:

> For the cross was not made of ordinary wood. A branch was first broken off from the tree of good and evil, then planted on the grave of the first man. This branch developed, grew with such magnificence that no tree could be compared with it. That was the predestined wood, the mysterious imperishable trunk. It was one of its branches which Noah's dove brought to the ark. Another, later on, furnished Moses with the staff with which he struck the rock. It furnished the bridge over the torrent proposed by Solomon to the Queen of Sheba and on which, through respect, she refused to set her foot when she entered Jerusalem. It was then miraculously preserved in the depths of the pool in Jerusalem until the moment of the Passion.[25]

It is remarkable that Claudel cites the passage with precision but without recalling its source. We find in this excerpt a condensed and finely articulated synopsis of the tree's trajectory from its roots in the Garden to its new form as crucifix. Claudel was certainly not the first to propose this connection between the fall of humanity and its redemption. One could say, in fact, that Claudel's assertions are consonant with Catholic dogma, reinforcing the inner logic of God's string of mysteries and the metaphysical connectedness of each biblical episode. The instability of Europe in Claudel's time was the source of a renewed spirituality in France, when neo-Catholicism found a weakened populace amenable to its message. During this interwar period after the Great War, in which France lost more lives than in World War II, it is clear that Claudel attempts to find meaning in the universe, to reroot the human in some sense of purpose. He was not the only one to turn to the metaphor of rootedness as a path to transcendence and a possible solution to Europe's disorientation in the first half of the century. As we will see in chapter 4, Simone Weil also found this metaphor useful in thinking about ways to restore Europe's former greatness after World War II, as did Martin Heidegger, who proposed a new kind of groundedness as the solution to humanity's distress in coping with new technologies in the atomic age.

25. Claudel, *Poet*, 46n34.

VALÉRY AND THE VEGETAL BRAIN

Like Claudel, Paul Valéry (1871–1945) invested the tree with symbolic power and vitalist purpose. In "Dialogue de l'arbre" (1943), Valéry stages an exchange between Lucretius, the first-century Roman philosopher-poet most famous for his *De rerum natura* (*On the Nature of Things*), and Tityrus, the shepherd figure from Virgil's *Eclogues*,[26] as the two reflect on a tree and all its intimations. As Theodore Ziolkowski notes, "Tityrus represents the poet who apprehends nature directly and instinctively while Lucretius, like Valéry himself, perceives meanings and relationships."[27] Without restraint, it is Tityrus who recounts his passion for the tree with spontaneity and starry-eyed wonder. His unmeasured meditations dwell on the surface of the tree, on its visible form, and on the reveries triggered by its image. In contrast, Lucretius puts himself in the place of the tree, embodying it and embedding his own subjectivity in it. He says to Tityrus:

> I wished to speak to you of the feeling I sometimes have of being Plant myself, a Plant that thinks, but that does not distinguish its diverse potencies, nor its form from its force, nor its port from its place. Forces, forms, size, and volume, and duration are but a single river of existence, a tide whose liquid expires in hard solidity, whilst the dim will of growth rises and bursts, and would again become will—in the light and innumerable form of seeds. And I feel myself live the unheard-of enterprise of the Type of the Plant, invading space, improvising a dream of branches, plunging in the midst of the mire, and drinking in the heady salts of the earth, whilst in the free air it opens by degrees to the bounty of the sky, green thousands of lips. . . . As much it goes down deep, so much it rises up: it chains the shapeless, it attacks the void; it struggles that it may change all into itself, for that is its Idea! . . . O Tityrus, it seems to me I am sharing with my whole being in that meditation—powerful, active, and rigorously followed up in its design—which the Plant bids me make.[28]

Again we encounter botanical purpose, here termed "dessein," comparable to Paul Claudel's image of the holy tree charged with a "devoir végétatif." Valéry's vitalist vision is apparent throughout his oeuvre, notably in well-known works

26. Valéry undertook a translation of Virgil's *Eclogues* from 1942 to 1944.

27. Ziolkowski, "Twentieth-Century," 159–60.

28. Valéry, *Dialogues*, 172.

such as his poem "The Graveyard by the Sea" ("Le cimetière marin") in which the poetic voice implores, "One must try to live" ("Il faut tenter de vivre"). Despite Valéry's denouncements of Bergsonian vitalism, critic Maurice Bémol identifies vitalist tendencies in his poetic oeuvre, calling him "un grand poète de l'élan vital."[29] As we will see, his character Monsieur Teste is the result of the vitalist pursuit of extending the life of an idea. In Valéry's "Dialogue," he depicts the tree not only as a life-form but as an organic being whose whole *raison d'être* is the expansion of itself as a body and as a *concept*; let us not forget that Saussure used the figure of the tree in his *Cours de linguistique générale* (1915) to illustrate the relationship between the signified, the signifier, and the referent. In Lucretius's monologue, knowing and being merge into a single concept: the philosopher's botanized body. The plant, with its "dessein," or purpose, impels him to put himself in its place, to live the visceral and the veneer simultaneously, to embody the tree and the idea of the tree. In general, the connection between the botanical and the expansion of knowledge is evident in concepts such as the *seminar*, originally a word used to describe a breeding ground for plants (*seminarium*) but adopted later through German to mean the student-professor milieu where learning is a kind of germination. Describing someone as *cultivé*, cultivated or cultured, also equates knowledge with botanical growth, as does the "planting of an idea" in someone's mind. This botanical and epistemological metaphor abounds in Valéry's oeuvre, in which he often organicizes the thought as a plant, but also as an animal or even a human in the case of Monsieur Teste.[30] The plant, then, is a figure for the expansion of thought and a metaphor for the way being organizes itself. In Lucretius's account, the plant "enchaîne l'informe, . . . attaque le vide," translating nothingness into being, but with fervency and purpose. In an aphorism from Valéry's "Analecta," he writes, "Consciousness is horrified by the void" (712), suggesting a shared need by the brain and the tree to fill the abyss with somethingness. The tree improvises a "rêve de ramure" or "dream of

29. Bémol, *Paul Valéry*, 426.

30. In his "Analecta" (1926), Valéry calls ideas his "little mental creatures" ("petites créatures mentales") and "seeds" ("germes"). Valéry, *Œuvres*, 700. This pattern recurs throughout the text: "There is an incalculable difference, an indeterminate interval; between the embryo of an idea and the intellectual entity it can finally become" (701); and "My kind of mind is not to learn books from beginning to end but to find in them only the seeds that I cultivate in myself, in isolation" (715). He claims that algebra gives "the impression of vegetable work, of a repetition that spreads out, one cell that is subdivided" (717).

branches," making the imagined real, pushing *puissance* ("potentiality") into *acte* ("actuality"). The tree is life; it is knowledge; it is being.[31]

In his dialogue with Tityrus, Lucretius also describes the tree as a kind of river, with its various ramifications resembling tributaries and its roots pushing forth through the dark soil with purpose: "There is no hideous beast of the sea more greedy and multiple than this tousle of roots, blindly certain in their progress towards the depths and dampnesses of the earth. But their advancement proceeds, irresistible, with a slowness which makes it implacable as time. Into the empire of the dead, of the mole and the worm, the toil of the tree inserts the powers of a strange subterranean will."[32] The plant's purposiveness is depicted here as "une étrange volonté souterraine," strange in its uncanny resemblance to human will. One immediately recognizes that Guillevic's understanding of vegetal will as explored in chapter 2 is quite sympathetic to Valéry's description of it. Moving downward and laterally, the plant improvises its own nutritive siphoning mechanism and drops its own organic anchor, securing a stable base for the ladder upward into the celestial sphere. This transcendence has little to do with God or spiritual transcendence; it is intellectual, a new frontier for the possibilities of the human mind.[33] Near the end of their dialogue, Lucretius says to Tityrus: "Is not to meditate to deepen oneself in Order? Just see how the blind Tree with its diverging limbs grows up about itself, faithful to Symmetry. Life in it calculates; it raises up a structure; and radiates its rhythm through branches and their twigs, and every twig its leaf, even at the very points marked by the nascent future."[34] Paul Valéry, an established literary voice in France in the first half of the twentieth century and, like Claudel, a member of the Académie française, constructed the tree as a figure for the proliferation of thought and being. In the examples of botanically metaphorized transcendence I have pointed out in this section, the rooted human has the potential to follow the plant as a model and surmount limits, often self-imposed, going beyond the state of the actual to the state of the ideal. As we shall see in the following pages, Valéry infuses another of his characters, Monsieur Teste, with this same botanical *puissance vitale* (vital potency or

31. For an analysis of the problem of source in Valéry's work, see Derrida, "Les sources de Valéry."

32. Valéry, *Dialogues*, 158.

33. A recurrent question for Valéry's Monsieur Teste is, "Of what is man capable?" ("Que peut un homme?"). Valéry, *Œuvres*, 23.

34. Valéry, *Dialogues*, 173–74.

power) and transcendent thrust shared between Lucretius and the Tree. Teste, however, is not an upright plant but an inverted one.

INVERSION AND CONVERSION

Thus I, easy philosopher,
Among the birds and trees confer:
And little now to make me, wants
Or of the fowle, or of the plants.
Give me but wings as they, and I
Straight floating on the air shall fly:
Or turn me but, and you shall see
I was but an inverted tree.

ANDREW MARVELL, "Upon Appleton House," 1651

The inversion of plants has a long history in the natural sciences, literature, and philosophy of the West. In the natural sciences, this inversion unfolds in the experimental mode as a trial meant to discover the adaptability of plants to new positions other than those in which they naturally orient themselves. Through such an experiment, the scientist interrogates the flow of the plant's sap and the water and minerals it ingests, questions the role of light and darkness and their effects on roots and branches, and asks whether a plant's orientation is absolute or relative. Must branches always grow away from the earth's center and the roots toward it, or does the plant have a natural calibration mechanism that helps it recognize up from down? In literature and philosophy, however, this type of inversion generally equates plant with human, necessarily acting as a figurative device to explore human adaptability. Similar experimental objectives still apply, however. Is the human's existence absolute or relative? Do humans really know up (transcendence) from down (banality at best, failure or damnation at worst), or are they so fixed in their patterns of existence and thought that they persist in one or the other direction out of habit rather than sureness of course? In their figurative iteration, uprooting and replanting mean revealing what was hidden and hiding what was earlier visible; they mean derailing a "natural" process to see how the human responds. There is a double violence in this action: first, a ripping of the roots from the soil, then a subsequent suffocation of the leaves and stems in the cavity that is left. One could imagine claustrophobia and agoraphobia as immediate responses, as the green and breathing top of the plant is buried in the asphyxiating soil, and the roots, so at ease in the dark and quiet, are suddenly exposed and naked to the light and the elements. The reversal is both a physical and a metaphysical upheaval.

In order for plant inversion to be read as a figure for metaphysical conversion, an analogy must be made between plant and human, a *topos* with a long history in European thought and literature. Rather than provide exhaustive examples of this analogy, I'll limit myself to a few cases pertinent to the connection I will make between (up-)rootedness and conversion. First, Aristotle argued in his *De partibus animalium* that the animal and the human are both a sort of inverted plant: "Thus the animal becomes a plant that has its upper parts downwards and its lower parts above. For in plants the roots are the equivalents of mouth and head, while the seed has an opposite significance, for it is produced above at the extremities of the twigs."[35] Plants, then, modeled an inverted version of perfection. Animals achieved the switched order of organs but not the verticality. Only humans were able to combine the animal's inverted bodily functions and the plant's verticality into the perfect divine being. The noble, vertical bearing of the humans was prefigured in the bearing of the plants.

The philosopher and physician Julien Offray de La Mettrie attempted to create an analogy between the human and the plant in his strange text *L'homme-plante*, published in 1748. In the preface, he writes, "Here, man is metamorphosed into plant, but don't assume that this is a fiction suited to Ovid's taste. The simple analogy of the botanical realm and the animal realm made me discover that the main components of one could be found in the other."[36] Both in Aristotle's depiction and in La Mettrie's, the human and the plant are considered not in their wholeness but in terms of the individual constituents of their bodily mechanism. For some thinkers and botanists, such as Aristotle and Edouard Grimard, the roots are analogically the head of the human.[37] For others, such as La Mettrie and the English agriculturalist Jethro Tull, the roots are the stomach or digestive system.[38] For still others, the roots are compared to stationary feet.[39] One could make equally compelling arguments

35. Aristotle, *On the Parts of Animals*, 117.

36. La Mettrie, *Œuvres philosophiques*, 50.

37. Grimard writes, "What is the root then? It is the head down below." Grimard, *La plante botanique simplifiée*, 80.

38. "And it is very likely and may be proved by another method, that as roots are but guts inverted, they do bear, perhaps, that proportion to the stem or stalks of plants, as guts do to the bodies of animals, viz. several times longer than the stalks." Tull, *Horse-Hoeing Husbandry*, 20.

39. In a letter from 1821 to Alfred de Vigny, Victor Hugo writes, "Why then do we have feet and not roots, if we are fixed like miserable plants to a spot that we cannot leave?" Hugo, *Correspondance*, 14.

for each of these positions, illustrating in the end that the human body is analogically comparable to that of the plant only when considered piecemeal, not as a whole.

In the early twentieth century, the American botanist Luther Burbank (1849–1926) attempted to think analogically about the plant and the human, not in terms of their anatomy but in terms of their capacity to crossbreed. In his 1907 text *The Training of the Human Plant*, Burbank writes:

> I have constantly been impressed with the similarity between the organization and development of plant and human life. While I have never lost sight of the principle of the survival of the fittest and all that it implies as an explanation of the development and progress of plant life, I have come to find in the crossing of species and in selection, wisely directed, a great and powerful instrument for the transformation of the vegetable kingdom along the lines that lead constantly upward. The crossing of species is to me paramount. . . . Before passing to the consideration of the adaptation of the principles of plant culture and improvement in a more or less modified form to the human being, let me lay emphasis on the opportunity now presented in the United States for observing and, if we are wise, aiding in what I think it fair to say is the grandest opportunity ever presented of developing the finest race the world has ever known out of the vast mingling of races brought here by immigration.[40]

This passage is noteworthy for several reasons. First, it is of interest to us in its depiction of selective breeding as an upwardly progressive phenomenon, a transcendence of a perceived genetic stagnancy in the human species. Burbank is essentially proposing an improvement of the human genetic pool by imitating the crossbreeding experiments he has found to be successful in plants. In other words, it is the botanical world that offers itself as a model to humans for the advancement of the species. His description of the trajectory of a species toward greater fitness as "constantly moving upward" signals his understanding of the evolutionary process as a transcendent one. Second, Burbank argues that the United States, given its "naturally" diverse immigrant population, provides perfect laboratorial conditions for an experiment that could confirm his hypothesis that species improvement is possible through "wisely directed" crossbreeding. Finally, Burbank seems to suggest that humans could be easily persuaded to offer themselves up for experimental breeding, taking on the same passive role as test subjects as the plants had. This would involve

40. Burbank, *Training*, 3–5.

a willingness on the part of the subject to breed with whomever the scientist deems most fit as a genetically compatible partner. Following the "voie végétale" or "botanical path" is the way to better humans.

While Burbank makes it clear that he is consciously applying his discoveries about plants to the human in order to initiate an improvement of the species, other botanists' experiments were only retroactively seen to have some relevance to the human, often as metaphorical and moralizing applications. In the eighteenth century, the discipline of botany thrived. Thousands of strange experiments on trees, flowers, mosses, grasses, and other forms of botanical life were conducted throughout Europe during the period, but one in particular has held its ground in French cultural memory. The botanist Henri Louis Duhamel du Monceau, who eventually conducted hundreds of experiments designed to test the adaptability of plants, had the peculiar idea of uprooting trees and replanting them upside down to see what would happen. The objective of Duhamel's plant inversion experiment was to confirm the directionality of sap flow and test the role of gravity in this flux. He doubted that the circulation of sap in the botanical bodies was analogous to the circulation of fluids in animal bodies. The centripetal vector of the plant's fluids in normal conditions made him wonder whether this centeredness was relative or absolute. His own account of these trials adopts the objective voice of the scientist. However, the imaginations of future writers such as Gaston Bachelard, Paul Valéry, Michel Tournier, and others borrowed this intriguing curiosity as a figure for human transformation.

Edouard Grimard provides a colorful account of Duhamel's tree-inversion experiment, in which the botanist tortures the plant by thrusting its breathing head underground but proves the plant's incredible improvisational power to reverse itself. While the author does wax poetic in his depiction of the experiment, his personification of the willow and his appeal to the reader to sympathize with the tree (what Francis Ponge might call "siding with things"[41]) provides another example of plant and human viewed as mutual surrogates. The critical responses to Duhamel's experiment reflected an aversion to the unorthodox: the scientist was reversing the order of things. This experiment in particular was often cited as an example of a scientific quirk. However, the Encyclopédistes praised Duhamel's experiment as the kind of unconventional trial necessary for the progression of secular humanism. In the tree entry

41. The title of Ponge's *Le parti pris des choses* (1942) could be translated as "siding with things." Given the year of its publication, it seems that taking the side of things rather than of humans is a sensible appeal. For Grimard's account, see Grimard, *La plante botanique simplifiée*.

(“*Arbre*”) of the *Dictionnaire raisonné universel d’histoire naturelle*, Pierre Daubenton and Denis Diderot chronicle the method of Duhamel’s experiment in an objective tone. However, they close the “Arbre” entry with the following passage:

> It follows that, from all the above examples, the more one studies nature, the more one is astonished to find, in the most seemingly base subjects, phenomena worthy of all of the attention and curiosity of the Philosopher. It is not enough to follow nature in its ordinary, settled course; one must try to throw it off course in order to know all of its fecundity and resources. The people will laugh at the Philosopher when they see him busy in his garden uprooting trees only to put their crowns in the ground and their roots in the air. But these people will be amazed when they see the branches take root and the roots become covered with leaves. Every day, the wise man plays the role of Democritus, and those around him, the role of the Abderites. This adventure dates from the first ages of Philosophy and continues today.[42]

This remarkable passage draws an analogy between the task of the philosopher and that of the botanist, or rather, it labels the botanist *as* a philosopher. Both objects of derision and doubt, these figures who embody the life of the curious mind do not rely indiscriminately on the methods of their forebears, nor are they satisfied to accept prevailing beliefs as indisputable fact. They are not deterred by implausibility and do not allow their neighbors’ skepticism to sway them from their unorthodox projects. Democritus, the laughing philosopher, is cast here as an antecedent of and a model for the eighteenth-century philosopher and the botanist willing to take risks. All three figures—Democritus, philosopher, and naturalist—are sage visionaries capable of seeing already what others cannot yet.

Diderot and Daubenton could have chosen among countless strange experiments to illustrate the tension between the persevering thinker and his sneering observers. Duhamel’s tree inversion, however, was particularly effective because it illustrates an inversion of nature and brings together, in the Platonic tradition, the human, the plant, and that “most authoritative part of our soul . . . living in the summit of the body.”[43] While for Plato, the head houses the soul, for Daubenton and Diderot, it houses the *esprit*, the “mind” rather than the “spirit.” Rhetorically, Daubenton and Diderot clearly attempt

42. Diderot and Daubenton, “Arbre,” 588.
43. Plato, *Timaeus*, 88.

to persuade the reader to consider a different way of seeing and being in the world and to understand that transformation is possible. Modeling the ideal enlightened mind, the Democritus-Duhamel analogy promotes risk and unorthodoxy. However, given the accusations that the Enlightenment itself is not immune to dogmatic proclamations, the transformation they propose could easily be read as a proselytizing effort, as a coaxing toward conversion. Their appeal, though indirect, urges the metamorphosis of a scoffing populace into a reflective, curious, and bold nation where reason trumps superstition or blind faith. The inverted plant is not only an object of curiosity for the incredulous onlookers; it is, by extension, a figure for how they should change their minds.[44]

MONSIEUR TESTE, BOTANICAL THINKER

Through the metaphor of the inverted plant, Paul Valéry casts his character Monsieur Teste as an exceptional brain, albeit of a very different sort from that of the philosopher-botanist depicted by Diderot and Daubenton. Valéry imagined knowledge to organize itself in the manner of a tree: "Knowledge extends like a tree, by a process identical to itself; by repeating itself. *Novat reiterando.* [It renews itself through repetition.]"[45] It is not surprising, then, that the tree is a central figure in Valéry's oeuvre, one so omnipresent that Pierre Laurette dedicated an entire book to parsing the tree imagery in the writer's corpus.[46] Nor is it surprising that Valéry would imbue his most cerebral character, Monsieur Teste, with botanical attributes. Valéry's "Teste cycle" was published first as a series beginning in the late nineteenth century and then as a complete collection in the early twentieth century. Monsieur Teste is essentially an anthropomorphized thought. Valéry even went so far as to call the book "the novel of a brain" ("le roman d'un cerveau").[47] For him, the challenge of Teste's

44. C. G. Jung offers a rich history of the inverted-tree image, including its presence in the Bhagavad Gita, in Dante's *Purgatoria* as the *arbor inversa,* in Andrea Alciati's *Emblemata cum commentaris,* and in other ancient and modern sources in the East and the West. See the chapter "The Inverted Tree" in Jung, "Philosophical Tree," 13:311–15.

45. "La connaissance s'étend comme un arbre, par un procédé identique à lui-même; en se répétant. Novat reiterando." Valéry, *Cahiers,* 3:273. Note: While "connaissance" can also mean "consciousness," "knowledge" seems a better translation in this case.

46. See Pierre Laurette, *Le thème de l'arbre chez Paul Valéry* (Paris: Klincksiek, 1967).

47. Valéry quoted by Maurice Toesca in Valéry, *Œuvres*, 1386. For more on Valéry's engagement with the brain in his writing, see Gabriele Fedrigo, *Valéry et le cerveau dans les "Cahiers"* (Paris: L'Harmattan, 2000).

existence is one of duration: how to extend the duration of a thought on a page, thus giving it life, cultivating the potential of the seed? Monsieur Teste, whose name could mean "head" or "witness,"[48] dedicated his existence to several tasks, the most important of which lies in his injunction "*Maturare!*" ("to ripen or mature" in the infinitive but with an implicit imperative entreaty), a vitalist notion depicting thought and body unfurling organically in time. Teste shows deep interest in human potential, particularly of the cognitive sort. The narrator who scrutinizes Teste's words and movements in an effort to decipher the mechanism of that monstrous brain explains, "This man had known quite early the importance of what might be called human *plasticity*. He had investigated its mechanics and its limits. How deeply he must have reflected on his own malleability!"[49] The question that haunts this singular figure the most is perhaps this: "Of what is man capable?" (23).

Among the texts of the Teste cycle, readers find a letter supposedly from Emilie, the wife of Monsieur Teste, to the friend who has narrated his experiences with the curious man in early texts like "La Soirée avec Monsieur Teste." I write "supposedly" because Valéry includes a curious prefatory note in the letter's first appearance in print, in the publication *Commerce* in the fall of 1924. This note is ostensibly written by the editors of the publication ("la Direction") and calls into question the authenticity of Emilie Teste's testimony about her husband: "We wondered whether this letter was perhaps forged by Mr. Teste himself, with some inexplicable intent."[50] The editors (or their fictional avatars as imagined by Valéry) are specifically hesitant in regard to the style of the letter ("It is hardly the language of a woman one finds in it")[51] and the idiosyncrasies of the writing, such as "a certain absence of restraint and modesty in the expression,"[52] all of which leads them to believe that Teste has ventriloquized his own wife, if she even exists at all. This "Note de la

48. Jacques Derrida cites "Pour un portrait de Monsieur Teste," which makes an explicit link between *testis* and *témoin* (witness). Derrida, "Les sources de Valéry," 593. Anne Mairesse also interprets the name to mean *witness*: "I would like to present the Valéry who, in my opinion, has been significant in the twentieth century, as a witness and testifier, as *Teste* (*testis, testa*) undoubtedly, but also, as I would qualify here, as 'the anthropologist.'" Mairesse, "Return to Monsieur Teste?," 1003. According to Michel Jarrety, the name Teste is "simply the name of a resident of Valvins whom Mallarmé met." Jarrety, *Paul Valéry*, 205.

49. Valéry, *Œuvres*, 11–12.

50. "Nous nous sommes demandé si cette lettre n'aurait pas été forgée par M. Teste en personne, dans un dessein qui ne s'explique pas?" Quoted in Valéry, *Œuvres*, 1393.

51. "Ce n'est guère le langage d'une femme qu'on y trouve." Quoted in ibid.

52. "Une certaine absence de retenue et de modestie dans l'expression." Quoted in ibid.

Direction" did not appear in later editions of the Teste cycle. It is implied that if the letter was written by Teste himself and not his wife, it is an opportunity for the man to communicate to his friend, the letter's addressee, his own (idealized) account of how he should be seen. The strange configuration of writing himself through another allows him to convey both his inner and outer features, those that are manifest to him and those perceived by others in a single portrait. He takes himself as an object of study, trying to adopt a point of view he could not possibly have.

Because the fictional editors were unsure of the letter's authenticity and this letter was left out of later editions of the Teste cycle, let's momentarily suspend our misgivings and trust that Emilie is in fact its fictional author. She compares her husband to "a mystic without God" (31) and notes, as so many others have, that the man's strange cerebral pursuits can be perceived but not understood by those who observe him. He must use his entire vital strength to pull himself from the depths of his mental digressions and restore contact with the extra-cerebral world.[53] The letter abounds with botanical imagery,[54] but one of these metaphors is particularly striking: Emilie describes her husband's soul as an inverted plant. She writes, "His soul, no doubt, changes into some peculiar plant whose root, and not the foliage, would thrust against nature toward the light!"[55] Valéry's Platonic reference is clear. However, given that Teste is compared to a mystic without God, the preoccupations of his soul are not bound to divine transcendence but to some other kind of transcendence, still tied to the *esprit* but to the cognitive rather than the spiritual kind. Throughout the Teste cycle, his freakish or monstrous nature is emphasized, suggesting that he is an inverted human, an aberration. An archaic term in both French and English for a homosexual was "invert" (*inverti*), one who goes against nature, *à rebours*. Teste, too, goes against nature, not bodily but cerebrally. Despite his

53. "He spins out in himself such fragile threads that they survive their delicacy only with the concerted help of all his vital powers." Ibid., 26.

54. "His rare tenderness is a winter's rose." Ibid., 24; "I tell myself that this man is perhaps damned, that in his company I am in great danger, and that I am living in the shade of an evil tree" (29); She describes their daily walk this way: "We would go down where you would like to go if you were here, to that ancient garden where all those who think, or worry, or talk to themselves, go down towards evening as water goes to the river, and gather necessarily together" (33). The scene and the letter end with Monsieur Teste's scoffing at the "ridiculous order" of the garden, with the plants neatly organized and labeled with plaques that list the names of each species. He says of the garden, "This is a garden of epithets . . . a dictionary and cemetery garden" (34).

55. Valéry, *Œuvres*, 29–30.

efforts toward a higher order of cognition, he still seeks groundedness, in an alternation between testing the limits of human consciousness and returning to his wife, who provides a solid ground after these mental divagations and who represents the world, the Earth Mother. As a transposition of the typical human, whose base corporeal needs take precedence over mental life, Teste is a textual model for how to do one thing in particular: "Penser autrement."

TOURNIER AND THE UPENDING OF WESTERN CULTURE

Inspired by his mentor and former teacher Paul Valéry, who had taken interest in the plant as a figure for the life of the mind, Michel Tournier (1924–) uses the inverted plant as a figure for metaphysical transformation in his novel *Friday* (*Vendredi, ou les limbes du Pacifique*, 1967), which won the Grand prix du roman de l'Académie française. Tournier was also a student of Gaston Bachelard, who wrote extensively on the subconscious associations expressed through the figure of the root, as outlined in chapter 1. With these precedents in mind and with a solid background in philosophy, Tournier set out to model the change he believed possible in the European mind, using a canonical European literary figure to do so. His retelling of Defoe's *Robinson Crusoe* recounts the shipwreck and arrival of the protagonist Robinson on a deserted island off the coast of Chile, an island he later names Speranza. The tale alternates between third- and first-person narration in the form of a log-book of Robinson's reflections about his experiences on the island. The log-book as reflective venue is borrowed directly from Valéry's Teste, who recorded his thoughts in the same format. Because Tournier had studied philosophy before turning to writing many years later, heavily philosophical concepts are omnipresent in the narrative, particularly in Robinson's meditations.[56] In the early stages of his arrival, he attempts to "civilize" the island despite the lack of other people on it, his only companion being a dog that had survived the shipwreck. He creates, for example, a charter for the island and a penal code. Furthermore, he repeats in a compressed time span a series of civilizational developments:

56. Tournier, a classmate of Gilles Deleuze at the Sorbonne, failed his *agrégation* in 1949 and abandoned the vocation of philosophy for a career in radio. Tournier's references to the ideas of thinkers such as Spinoza, Leibniz, Kant, Sartre, Hegel, and many others are generally quite obvious or explicitly stated in subsequently published interviews and essays about his work. For more on Tournier's use of philosophy, see Colin Davis, *Michel Tournier, Philosophy and Fiction* (Oxford: Clarendon Press, 1988).

history (he keeps a log-book), cartography (he makes a map), time-keeping (he keeps a calendar and builds a clypsedra, or water clock), religion (he reads the Bible and performs various Christian rituals). The narrator explains, "After cultivating the soil of the island, he had now given it the beginnings of a dairy herd. Like mankind at the dawn of history, he had passed through the stage of hunting and gathering into that of tilling and stock raising."[57] In short, he becomes the administrator of the island. This administrative obsession protects him from what he perceives as the chaos of the island and its nature. However, the discrepancy he notices between "l'île administrée" and "l'*autre île*"[58] foretells his conversion to come:

> For if on the surface of the island I pursue the work of civilization—tillage, stockbreeding, building, administering, lawgiving—which follows the pattern of human society and is therfore in a sense retrospective, I feel that in myself I am the scene of a more radical process of creation, one which is engaged in finding new and original substitutes for the ruins that solitude has left with me, all more or less tentative and so to speak experimental, but bearing less and less resemblance to the human model from which they sprang.[59]

It is essential to note Robinson's word choice in the expressions "solutions originales" and "l'évolution plus radicale" in this excerpt from his log-book. In contemporary contexts of theory and politics, the connection between "radical" and "root" is often elided. Does radicality involve an acknowledgment and appreciation of an original state and the celebration of a return to it? Or does it involve the total eradication of an institution and a subsequent replanting of it? Is radical simply a synonym for extreme, fundamental, essential, total? Why is radicality most often associated with Leftist movements? Because it is the opposite of conservatism, the will to preserve? In Tournier's case, the term suggests a total overhaul of his protagonist's existential habits. Robinson's transformation throughout the novel involves a return to what is fundamental, what is *élémentaire*, an adjective used in the book to signify a reconfiguration of the basic units of existence and the conversion from a telluric to a solar being.

Robinson's first contact with humans after the shipwreck occurs when Araucanian Indians from Chile arrive on Speranza in order to conduct ritual sacrifices. Although Tournier does not use it, the word *autochtone* is often

57. Tournier, *Friday*, 49.
58. Tournier, *Vendredi*, 140, italics in original.
59. Tournier, *Friday*, 111.

used in French to designate indigenous people. The word can be traced to the Greek root *khthon* (earth) and means "one sprung from the soil he or she inhabits." This etymology points to an analogy between the plant and the human as entities born from the soil. The tree is often conceived as a bridge between the Chthonic (underworld) and the Uranic (celestial sphere), and Tournier incorporated these notions directly into his novel. When the Auracanians return for another ritual, one of them is spontaneously selected for sacrifice, one who "was also darker-skinned and somewhat Negroid in feature, in general different from the other men" (134). When Robinson's dog Tenn, his sole companion, barks, thus drawing attention to Robinson from his hiding place, he decides that the best course of action is to shoot the designated sacrifice victim, who had managed to escape during the distraction caused by the bark. The dog deviates Robinson's aim, causing him to kill one of the pursuants instead. From that moment on, Robinson has a new human companion—the man who narrowly escaped being sacrificed. He gives him the name Vendredi (Friday), after the day on which he saved him.

Vendredi is the conduit by which Robinson's conversion is made possible. Through a series of initiations, Robinson moves from the earthly to the solar, in essence emptying his head stuffed "with the trappings of three thousand years of Western civilization" (135). At the end of his conversion, he recognizes Vendredi's role, writing in his log-book: "Friday forced me to make a more radical conversion,"[60] one that required him to couple literally with the island Speranza, making love to the soil and the plants as though they were the sex of a woman and descending deep into her womb, a cave from which he is eventually reborn. He later called this timeless period of descent into the cave "the earth-bound phase of his life" and his "his long sojourn underground."[61] In his time in the depths of the earth, Speranza transforms into mother, recalling the primordial origins of humans, having sprung from the earth. Robinson even comes to understand that the vertigo produced in humans when they climb trees or cliffs is a result of the terrestrial nature of the human, who is stubbornly bound to its element of origin.[62] In his log-book, he writes of the experience of

60. Norman Denny translates the sentence as follows: "It was Friday who brought about the deeper change." Ibid., 212. However, this rendering misses the significant formulation of "une conversion plus radicale" in Tournier's original: "Vendredi m'a contraint à une conversion plus radicale." Tournier, *Vendredi*, 229.

61. Tournier, *Friday*, 98, 105.

62. "He realized that vertigo is nothing but terrestrial magnetism acting upon the spirit of a man, who is the creature of earth. The soul yearns for that foothold of clay or granite, slate

quitting the cave: "It is written that we cannot enter the Kingdom of Heaven except with the humility of a child. Never was any word of the Scriptures more literally true. The cave not only affords me the unshakable rock on which I may rest my poor life; it is a return to the lost innocence which all men secretly mourn. It unites as by a miracle the peace of the sweet maternal darkness and the peace of the grave, the one before, the other after life" (107). Again, transcendence can be accessed only through the rootedness of self. As Anthony Steinbock writes, "It is by virtue of the Earth's function as ground that the lived-body can be self-giving and such that transcendent objects can be given at all."[63] Here, in a nostalgia for the matrix, Robinson reroots himself in the original soil, returning to an embryonic state. Only this telluric embeddedness, which he believes to be confirmed by the Gospel, can create the conditions for his vertical ascent and solar metamorphosis.

After this revelation, he embarks on a different educative path, what he eventually calls the *voie végétale*, the botanical way or path.[64] The narrator reflects, "Was [Robinson] the last member of the human race to return to the vegetative sources of life?"[65] In an evolving series of personified depictions of plants and trees, he begins to find them sexually attractive. After coupling for several months with a fallen *quillai* or soapbark tree, he is bitten one day by a spider while in the act and interprets this as a symbolic venereal disease and a sign that "the 'vegetable way' might be no more than a dangerous blind alley" (116). Despite the warning, he pursues his botanical intercourse. In his explorations on the island, he discovers a pink valley or combe, *une combe rose*, that attracts him.[66] He begins to reflect on the word *combe*, which triggers another word, *lombes*, or "loins," in his mind, the pink valley thus transforming as both image and word into a sexualized sign. Readers might also make the connection between *lombes* and the *limbes* of the title, which means "limbo" or the plural "limbs." In a complex set of associations between love (Eros) and death (Thanatos), Robinson decides that both are of an earthly nature ("de nature tellurique" [133]), because the earth welcomes lovers falling together in passion and the bodies of the dead. Robinson recounts his coupling with

---

or silica, whose distance at once terrifies and attracts, since it harbors the peace of death. It is not the emptiness of space that induces vertigo, but the enticing fullness of the earthly depths." Ibid., 189.

63. Steinbock, *Phenomenology*, 16.

64. Tournier, *Vendredi*, 121.

65. Tournier, *Friday*, 116.

66. Tournier, *Vendredi*, 127.

the ground: "I dug my grave with my sex and died the transient death we call pleasure. And I note that in doing so I have accomplished a further stage in the metamorphosis I am undergoing. . . . [I]n the pink coomb my sexuality returned to its original source, the earth."[67] After coupling with the earth, he notices that a new species of plant has sprung up exactly where he left his seed. He gently unearths one of the plants to discover the following: "It was true! His love-making with Speranza was not sterile. The white, fleshy, curiously forked root bore an undeniable resemblance of the body of a woman-child" (129). This mandrake child does not produce fertility, like the examples seen in chapter 1; it is rather the *result* of fertility. He notices furthermore that after his night of sleep lying on the ground, his beard had begun to take root in the soil.

His conversion is most clearly portrayed in the figure of the inverted plant. Friday disappears for a couple of days, and, out of concern, Robinson sets out to find him. He happens upon a "secret world," a kind of ritual space composed of masks, snakeskins, bodies of dead birds, and other objects. Friday has replanted plants upside down and disguised himself as a plant. Through the metaphor of these inverted plants, Friday foretells Robinson's conversion: an abandonment of the European values that had birthed him and the adoption of a more radical, elementary system of values. Upon its publication, Tournier's novel resonated with the public for reasons beyond his literary prowess. The text was well received because of its absolute relevance to the sociopolitical climate of the time. By the 1960s, France had lost most of its colonies, and the turmoil of decolonization destabilized long-held beliefs of the supremacy of the *métropole* in relation to the extrahexagonal peoples it had governed. In addition, in its effort to replenish a depleted workforce, France sought out a new working class that it might import from the colonies to keep the nation competitive. This measure had been conceived as a temporary solution; after the immigrants helped France to recover economically, it was presumed that they would somehow simply vanish, returning happily to the place whence they'd come. Tournier's novel obliquely addresses these predicaments, which had not yet found a productive articulation in public discourse. By recrafting Defoe's tale of that most European of Europeans, Robinson Crusoe—with his mercantilist ambition, his Christian certainty, and his administrative compulsion—Tournier clearly attempts to compensate through literature for the injustices dealt by Europe to its colonies. His novel certainly would have had less impact had he invented completely new characters without a long,

67. Tournier, *Friday*, 125–26.

influential cultural presence. Other than Montaigne's *Essais*, Defoe's *Crusoe* is perhaps one of the only texts that has spawned its own prodigious genre, the robinsonade. In Tournier's book, the Other as embodied by Friday is humanized in an unprecedented way. His influence on Crusoe is superior to that of Crusoe on him. He imposes his reality on Crusoe and the (French) reader. No longer a virtual, imagined subject across the globe, exoticized, zoomorphized, infantilized, reduced, and belittled, Friday is at your doorstep, urging to be dealt with through his mere presence. One possible answer to such an imposition of the real is total and deliberate conversion, a reconsideration of the universality of European values.

When Crusoe comes across Friday's shrine, he is shocked to find that nature has accepted the inversion imposed upon it. Because Friday has made his own body into that of a plant, he illustrates even more clearly to Crusoe that the inverted plants are not simply figures for themselves but figures for human convertibility. By analogizing his own body as plant, he makes clear that the human is a changeable creature. While the European had convinced himself that the only path to transcendence—specifically Christian transcendence—involved the strict observance of principles and commandments clearly given by the Scripture and predicated on the superiority of his race, suddenly a very different kind of transcendence is made available to him. Still vertical in nature, since his elemental conversion leads him upward toward the sun, this new iteration of transcendence is compatible with the lives of Others. Without the self-imposed Eurocentric white man's burden, Robinson finds a sudden lightness as he ascends toward his solar future. Tournier is essentially turning European culture back on itself through a maneuver of inversion. By playing out an alternate course of events in the hypothetical space of fiction, he undermines the European values extolled by Defoe's canonical realist novel. Rather than making his point with his own narrative and characters, Tournier found a much more effective conduit for a collective *prise de conscience* by tipping on its head an established literary artifact with a centuries-long resonance.

*

In these examples from Aristotle to Tournier and in disciplines such as natural science, philosophy, and literature, the inversions most often imply or make specific an analogy between human and plant. That one is taken for the other is a commonplace since antiquity. I have shown how the same figure in various fields confirms that, while transcendence is predicated on rootedness, the human does have the capacity to invert what was accepted as the proper course of nature in an improvisational adaptation to unforeseen change. In some instances, this analogy takes the form of a mechanistic comparison of bodily functions: the plant's roots are like the human feet, stomach, or head. In other cases, the human and the plant are considered more holistically, both having an ingrained vital will that accounts for their survival and growth. Even the most catastrophic event, a reversal of ground and sky, can be survived and perhaps welcomed as an occasion for positive change. The inverted-plant figure is often associated with the reflective life of the human. Because the proliferation of knowledge is frequently depicted as plant growth, the inverted plant can mean a change of mind. For the Encyclopédistes, it stands as a motif for the kinds of exceptional philosophical experiments that should be undertaken in the service of secular humanism. To ponder the previously imponderable is the task of the good thinker. Inversion for the thinker is not a violence; it is a pedagogical necessity for the spread of radical thought. That inverted plant-man Monsieur Teste is another who thinks "autrement," subverting social codes and the norms of how one is permitted to think. This anomaly thrives with his roots in the air, a curiosity to anyone who comes in contact with him. But as a witness (*testis*) to the other side, his self-imposed task of intellectual maturation leaves him exiled from the sphere of the social. After all, his entourage and those strangers with whom he comes into contact in the garden and on the streets have only seen one side of things. They are embedded in their way of knowing. Of all the examples, the story of Robinson-as-convert is the most provocative and politically relevant to its contemporaries. While the Encyclopédistes were certainly interested in upending the dominant patterns of thought among their compatriots, Tournier pits European culture against itself by reversing an endorsed canonical tale of European superiority. The inversion of the plants is a miniature tableau of what the entire novel attempts to do: flip the system of value to show its arbitrariness and unsustainability as grounded truth. Through metaphors of inversion and conversion, Tournier illustrates the very concrete problems touching the book's first readers. Unstable soil precludes both rootedness and transcendence.

As I have shown in this chapter, rootedness was a notion that first attached the human to the divine, then to the earth. In the early depictions cited, particularly Plato's, the soul of the human was automatically rooted in the divine, by birth and by nature. The ecocentric paganism of antiquity was compatible with this divine rootedness; Man had not yet "fallen." However, in the Christian imagination, the celestial is no longer a given for the human; rootedness to the earth was the condition into which Man was born, and his life was spent in the service of an active transcendence of these terrestrial roots.

Paradise was no longer granted by birth; it was to be earned. Furthermore, during the progressive sedentarization of the human, from the early moments of agriculture-based subsistence to the coalescence of the nation-state, the soil gained new symbolically charged connotations as the ground into which people could root themselves. This progression marks the increasingly deep schism between the *nomades* and the *sédentaires*, between rooted populations and unrooted ones. In the next chapter, we see how the connection between rootedness and transcendence presented dire consequences for those, particularly immigrants and Jews, who were perceived to be unrooted and unrootable in the national soil. In France, the prerequisite for acceptance among the proud descendants of "nos ancêtres les Gaulois" was a condition that could never be met: a demonstrable genetic, geographic, and cultural heritage of *francité* that could be neither bought, stolen, nor earned. And in Germany, where *Blut* was bound to *Boden*, proof of rootedness became the basic condition for survival.

CHAPTER 4

Saving Europe from Itself: Weil's *Enracinement* and Heidegger's *Bodenständigkeit*

Men are not trees; a metaphor is not an argument.

RÉMY DE GOURMONT, "Les transplantés"

As World War II showed the magnitude of its violence, two thinkers, one from France and one from Germany, attempted to draft a vision of how the wrongs of Europe could be rectified. Out of disillusionment, they turned to figurative language. And both authors chose metaphors depicting the human as plant: the first was Simone Weil's notion of rootedness or "en-rootment" (*enracinement*) and the second, Martin Heidegger's notion of groundedness (*Bodenständigkeit*). The philosopher and factory worker Weil argues in her book *The Need for Roots* (*L'enracinement*, published posthumously in 1949) that World War II was caused by a collective loss of roots. To have roots, she explains, means to participate in a given community, to actively venerate the treasures of the past, and to become a steward of the future. In her view, the decay of moral, intellectual, and spiritual roots may be caused in part by military conquest, the obsession with money, and an education system that disregards Europe's forebears the Greeks. Her Marxism and Christian mysticism are manifest in her insistence on a personal, spiritual investment in physical labor, an essential step toward the regrowth of roots. Weil's example clearly reinforces the connection between rootedness and transcendence established in chapter 3. She literalizes the root metaphor by emphasizing the importance of land ownership, not in the nationalist sense—to own specifically French soil—but as an obligation and privilege to become a custodian and involved resident of the earth. I argue that Weil's biggest fear was not war, Nazism, urbanization, industrialization, or a collective drift away from God; her supreme fear was abstraction. In Weil's vision, to be uprooted meant to live in a world guided increasingly by abstract principles.

For Heidegger, on the other hand, his preoccupation with groundedness could not be more politically identarian. As I was writing this book, the new volume of his *Black Notebooks* was published, refueling the debate over the extent of his anti-Semitism and to what degree it had informed his philosophical thought. The blatantly anti-Semitic passages in the third volume have made it impossible to claim that Heidegger was a victim of circumstance or that his philosophy was not heavily influenced by his politics. In his 1955 text *Gelassenheit*, he explores and reframes the notion of *Bodenständigkeit*, translated often as groundedness or rootedness, which confirms again the link between stability down below and transcendence up above in Western thought.[1] While his earlier obsession with *Heimat* (homeland) and regionally specific groundedness had the obvious nationalist overtones of right-wing (eco-)politics, his reformulation of the notion attempts to address the new reality of the modern age, namely that many Germans fled during and after the war and those who remained were in a sense always drawn elsewhere by television and radio. He saw technology as inimical to a sense of groundedness in the world. In this chapter, I show how Heidegger's appeal for the human to be put back in its place follows the same logic as the French concept of *terroir*, which combines agricultural science and regional *chauvinisme*. The belief that land leaves its signature on agricultural products and on people becomes a philosophical justification for entitlement to a specific parcel of land. The terrestrial nostalgia involved in *terroir* follows the same logic as the regionalist effusions of Maurice Barrès's *Les déracinés*. Heidegger, Weil, and Barrès rely on a restorative poetics, conveying through retrophilic language that the better past is recuperable.

TALK OF ROOTS IN THE AIR: *LA QUERELLE DU PEUPLIER*

Man . . . is not a tree, and humanity is not a forest.

EMMANUEL LEVINAS, *Difficult Freedom*

1. Robert Savage, translator of Hans Blumenberg's *Paradigms for a Metaphorology*, notes the link between Blumenberg's language of root and ground and that of Heidegger: "Blumenberg's contention that metaphors are historical 'in a more radical sense' than concepts, that they point the way toward an occluded, subterranean realm where the basic questions posed in the 'ground of our existence' (*Daseinsgrund*) have already been decided, resonates unmistakably with another contemporary type of theory design: the late philosophy of Martin Heidegger, whose influence on West German philosophy in the 1950s it would be difficult to overestimate." Savage in Blumenberg, *Paradigms,* 139.

It is no coincidence that Weil and Heidegger turned to similar metaphors as they tried to save Europe from itself. That two thinkers of such different backgrounds—one a Parisian woman of Jewish ancestry who converted to Christianity and was raised in an affluent family, and the other a man from a Catholic, lower-middle-class family in rural Germany who was complicit with National Socialism—would arrive at similar metaphors as the solution to Europe's woes attests to the ubiquity of the human-as-rooted-being trope during the period. Roots were in the air. In Germany, the term *Volksgemeinschaft*, or "people's community," was used as the organizing slogan during World War I to give a sense of *Verwurzelung* (rootedness) in the *Vaterland* and was later taken up by the National Socialists. In France, colonialism and the rhetoric of French universalism thrived under the Third Republic, despite the tense nationalist debates of the period.[2] This overlap of identity politics and territory disputes kept root language at the surface of public discourse. The incident in which the root made its most spectacular appearance was the early-twentieth-century debate between André Gide and Maurice Barrès known as the *Querelle du peuplier* (the Poplar Quarrel). This battle royale between the *nomades* and the *sédentaires* was the extension of the age-old European dispute about who counts as rooted.[3] Given the fallout from the Dreyfus Affair and the ceaseless polemics surrounding the political exclusion of the so-called Wandering Jew, the debate between these two writers became a public outlet for anti-Semitic opinions, in both thinly veiled and conspicuous formats. The argument began in 1897 with Gide's critique of Barrès's just-published *roman*

2. Naomi Schor writes, "Throughout the nineteenth century, despite the emergence of the countervailing forces of nationalism, republican revolutionary ideology became more firmly implanted, reaching its apogee under the regime of the Third Republic, which, not coincidentally, marked the apex of French colonialism. The Third Republic was, so to speak, the golden age of French universalism." Schor, "Crisis," 47. Ernest Renan's essay "What Is a Nation?" (1882) proposed the notion of subjective nationality, in which a person who feels aligned with the values of a certain nation elects it as her own. More recently, Alain Finkielkraut has taken up this concept in his project *Qu'est-ce que la France?*, a series of interviews he conducted with French thinkers on topics such as being black in France, problems of immigrant integration, forms of secularization, and Catholicism.

3. The conflict between the *nomades* and the *sédentaires* is one of the most repeated themes in the work of the French writer Michel Tournier, explored in *Eléazar ou la source et le buisson,* in "L'aire du muguet" in *Le coq de bruyère*, and in the essay "Le nomade et le sédentaire" in *Le miroir des idées*. In this last book, he writes of the "countless versions of this never-ending battle between nomads and sedentaries" (Tournier, *Mirror*, 37), including sedentary Cain's murder of nomadic Abel, the territorialization of America, the Saharan Tuareg and their conflicts with settled peoples, and the Nazis' demonization of the Jews.

à thèse The Uprooted (*Les déracinés*, 1897), dedicated to Paul Bourget[4] and the first in a series of three novels under the name *The Novel of National Energy (Le roman de l'énergie nationale)*.[5] His *Déracinés* argued in no uncertain terms that young Frenchmen from the countryside should avoid cities and should develop and nourish roots in their regions of origin. Barrès's novel, influenced by Hippolyte Taine's theoretical take on rootedness and Guy de Maupassant's literary one,[6] dramatized the regionalist celebration of his precious Lorraine. Uprooted people—namely, Jews, foreigners, and cosmopolitan intellectuals—were enemies of regional, cultural embeddedness. Emily Apter sees in the debate an emphasis on "the extent to which determinations of modern identity stubbornly revert to literal figurations of the botanics of roots and the physicality of land."[7] Barrès, a *boulangiste* and anti-Dreyfusard, spent his entire public life rallying the nation through his fiction, essays, journalism, and heavy political involvement. As Sarah Hammerschlag has argued, "nationalism, not Catholicism, served in the role of religion for Barrès."[8] Gide, in contrast, argued that extensive travel, which constitutes a special kind of learning, could provide for the strong character an occasion to increase the originality of thinking or acting.[9] Encouraging weak-willed people to stay at home, Gide wrote bitingly: "Leave rootedness to the weak ones; let them stay encrusted in their hereditary habits, which will keep them from getting cold."[10]

This well-documented and thoroughly analyzed debate, which unfolded as a ridiculous exchange of reviews and open letters informed by the research of sylviculturists, gardeners, and botanists, spun into utter silliness, with humans

4. As the epigraph to a chapter called *Mr. Taine's Tree*, Barrès cites Bourget's *Conversations:* "M. Taine, sur la fin de sa vie, avait coutume chaque jour de visiter un arbre au square des Invalides et de l'admirer." Barrès, *Déracinés*, 202.

5. The two other novels in the trilogy were *L'appel au soldat* (1900) and *Leurs figures* (1902).

6. In his *Introduction à l'histoire de la littérature anglaise*, Hippolyte Taine argues that three factors are responsible for what he calls the "elementary moral state"; these are race, place (*milieu*), and historical moment. He defines race as the "hereditary and innate dispositions" that mark both "the temperament and the body's structure." Taine, *Introduction*, 17. For more on Maupassant's influence on Barrès, see Jean Canu, "Maupassant et *Les déracinés*," *Revue d'histoire littéraire de la France* 41, no. 3 (1934): 448–51.

7. Apter, *Continental Drift*, 34.

8. Hammerschlag, *Figural Jew*, 34.

9. Hélène Cixous describes herself as a rootless writer, but because of her diasporic condition, not, like Gide, because she could not pin home to a particular region in France. She writes: "Je n'ai pas de racines à quelles sources pourrais-je prendre de quoi nourrir un texte. Effet de diaspora. Je n'ai pas de langue légitime." Cixous, *Entre l'écriture*, 24.

10. Gide, *Essais critiques*, 7.

being compared to poplar trees, lettuces, roses, cattle, and a seal.[11] Experts in botany and zoology piped up to clarify whether a better analogy could be made between the human and this species or that one. In an atmosphere of entrenched positivism, science came to the aid of the participants' relentless literalization of the metaphor. Scientists clarified many specialized terms, and the argument shifted to a semantic one with dissension over the glossing of words like *transplantation*, *dépaysement*, *déracinement*, *repiquage*, and *aération*. Perhaps the most excessive part was the avalanche of autobiographical details that many of the authors felt the need to disclose in order to bolster their claims, divulging details of where they were born, where their parents, grandparents, and great-grandparents had lived, the lifelong *parcours* that led from this city to that, into the country and back.[12] By itemizing their ancestors and each place where they had lived, these writers mobilized their autobiographies as proof. Origin became argument. Each person took himself as a case study, evidence to support whatever happened to be the person's position in the debate. On the conservative side, staunch nationalist declarations abounded, like that

11. In his excellent annotated edition of Gide's *Essais critiques*, Pierre Masson explains in meticulous detail who the protagonists were, their positions, and how the debate unfolded. See Gide, *Essais critiques*, 952–55, 1007–9. Recent critical analyses of the debate include Sarah Hammerschlag's chapter "Roots, Rootlessness, and Fin de Siècle France," in *Figural Jew*, 25–67; Jean-Michel Wittman, "Gide, un 'anti-Maurras'?," in *Le maurrassisme et la culture: L'action française*, ed. Michel Leymarie and Jacques Prévotat (Villeneuve-d'Ascq: Presses universitaires du Septentrion, 2010), 99–110; Philippe Piedevache, "Barrès, ou le revers de l'histoire: La question des racines," in *Résistances à la modernité dans la littérature française de 1800 à nos jours*, ed. Christophe Ippolito (Paris: L'Harmattan, 2010), 211–28; Emily Apter's chapter "Uprooted Subjects: Barrès and the Politics of *Patrimoine*," in *Continental Drift*; and Hilary Hutchinson, "Gide and Barrès: Fifty Years of Protest," *Modern Language Review* 89, no. 4 (October 1994): 856–64.

12. Here is just one particularly colorful example among many:

> Produit en variation désordonné entre Picard, Champenois, Bourguignon, Parisien et Briard enserrés jusqu'à vingt ans, entre l'asphalte de Paris et un atmosphère surchargé de respiration et de relent d'égoût; j'aime tout ce qui ne me rappelle pas la ville triste et sévère; et un seul sourire du soleil d'hiver en Provence ou en Languedoc me dit plus que le formidable rire échauffé de toute une salle de spectacle dans la Ville Lumière, et je préfère l'architecture de mon ramier de carolin à celle de l'Opéra. Alors, me voilà aussi, phoque pathétique, destiné à prendre place auprès de M. André Gide, et pourtant, n'en déplaise à Maurras, je suis de bonne foi. J'ai cherché à me raciner, sans résultat.
>
> Rouart, "Un prétexte," 261

of Charles Maurras, who wrote, "The homeland [*la patrie*] is a point of view, a constant one."[13]

Ihab Hassan, Emily Apter, and others have described the twentieth century as a shift from a root-based epistemology to a rhizomatic one,[14] and it is clear that in the *Querelle du peuplier* Gide's position already hinted at the coming rhizomatic age. His celebration of nomadism was glossed in the late twentieth century by thinkers who tried to conflate rootedness and rootlessness into a unified concept: rootedness in rootlessness. Several applications of this concept—including Jacques Derrida's *destinerrance*, Edouard Glissant's *errance enracinée*, Jean-Claude Charles's *enracinerrance*, and Vilém Flusser's "taking up residence in homelessness"—express a will to reconcile the polarities that turn migration and dwelling into politically charged acts. While Derrida's term, which fuses the words *destin* or *destinée* (destiny or fate) and *errance* (wandering), ostensibly involves the communicative deficiency that keeps a message from getting where it's going, the idea can be extended to describe the constant communicative distance that separates one person or one culture from another. Martin Jay calls it "the inevitable process of miscommunication."[15] In what most closely resembles a definition, Derrida describes *destinerrance* as "the possibility for a gesture not to arrive at its destination."[16] J. Hillis Miller calls *destinerrance* "a spatial figure for time" and writes that the term "names a fatal possibility of erring by not reaching a predefined temporal goal in terms of wandering away from a predefined spatial goal. *Destinerrance* is like a loose thread in a tangled skein that turns out to lead to the whole ball of yarn."[17] This string theory is as fuzzy as Derrida's concept. Miller, who scoured Derrida's corpus trying (with little success) to find a clear definition and the original use of *destinerrance*, expresses reservations about the term and his attempt to define it; he is aware of the traps Derrida sets for readers and of the unintended straying from what was meant. He

13. "La patrie est un point de vue. Un point de vue constant." Maurras, "La querelle," 2.

14. Hassan, in his grid of the differences between modernism and postmodernism, places the root on the side of the former and the rhizome on the side of the latter. Hassan, *Dismemberment*, 268. Apter writes that the early-twentieth-century debate between Maurice Barrès and André Gide over whether humans are rooted creatures "may be read as an early version of the rhizome's contest with the *racine*." Apter, *Continental Drift*, 33. See also Alicia Juarrero, "From Modern Roots to Postmodern Rhizomes," *Diogenes* 163 (1993): 27–43.

15. Jay, "Still Waiting," 28.

16. Derrida, *Sur parole*, 53.

17. Miller, "Derrida's *Destinerrance*," 893.

proceeds anyway, defining *destinerrance* as the conditions of possibility that "any utterance or writing I make may escape my intentions both as to what it should mean (for others), and as to the destination it is supposed to reach. I may be destined to err and to wander, even though it may sometimes, by a happy accident, reach the destination intended for it" (896–97). Miller points toward the important etymological link between the word "error" and the notion of straying (*errare* in Latin means both "to stray" and "to err"). His reasoning leads him to conclude that there is a measure of joy in the near misses of communication, that the destiny of a message (or a person) to wander, never fully reaching a recipient or a destination, is not cause for anxiety but an occasion for "playful exuberance."[18] In the case of textual *destinerrance*, thematic coherency is replaced by digression; logic is resisted; the root is rhizomatized. Already in Derrida's half-hearted theorization of the concept, one perceives his preemptive certainty that it won't reach the audience in the way he intended. Why bother to give a solid and stable form to a concept that is bound to liquefy on contact with the receiver or that will never make it to her at all? The notion of the self as a message that is never received, that is misinterpreted, or that is only partially comprehended communicates an anxiety about the gaps that separate one human from the next. All of the space that partitions one person from another contains within it the possibility for endless deferrals of meaning. The wandering destiny of the message, encoded in a body, goes astray. Miller is right to note the autobiographical inflections of Derrida's writing on *destinerrance*; the philosopher uses his theorization to depressurize a nervousness about incommunicability. A minor adjustment in thinking about messages that move erratically transforms anxiety into joy.

Glissant, in his reflections on *errance enracinée*, turns to the poet Saint-John Perse (Alexis Leger), the French national born in Guadeloupe who won the Nobel Prize in 1960. Glissant argues that Saint-John Perse idealized European poetry and placed himself in a European cultural genealogy rather than a Caribbean one. For him, poetry found its source "in an idea, in a desire, not in the literal fact of the birth."[19] The tensions between *ici* (here, the destination of

18. "Derrida, I conclude does not just name *destinerrance* as an objective fact. Nor does he just exemplify it in the local style and overall structure of his essays. He is also himself the joyfully willing victim, as one might call it, of *destinerrance.* A playful exuberance, or joyful wisdom, a *fröhliche Wissenschaft*, is an evident feature of Derrida's writings. However hard he tries to stick to the point, he is destined to wander." Ibid., 900.

19. "La poésie prend source dans une idée, dans un vouloir, non pas dans la littéralité de la naissance." Glissant, *Poétique*, 49.

Europe for which he strived) and *ailleurs* (elsewhere, the Caribbean whence he came)—paradoxical in Glissant's terms, since the "here" is typically where you are and the "elsewhere" is the place you could go—are felt in every page of Saint-John Perse's poetry. He manages to convey a universality through his rootedness in wandering, suggesting that individuals can free themselves from history and home and gain autonomy through their chosen name, a symbol of breaking ties with the circumstances life has imposed. Glissant argues that Saint-John Perse's poetry "heralds a new relation with the Other, which, paradoxically and precisely because of this passion for wandering, foretells the poetics of Relation."[20] Saint-John Perse's rootedness in errantry makes him the forerunner of the late-twentieth-century gravitation toward *Créolité* and archipelagic and relational thinking in the Caribbean.

Jean-Claude Charles (1949–2008), a Haitian writer who lived much of his life between New York and Paris, introduced the idea of *enracinerrance*, which contains both rootedness (*enracinement*) and wandering (*errance*), in a book published in 1980, the same year as *A Thousand Plateaus*, called *Le corps noir* (The black body).[21] Coincidentally, this text impeccably performs the same rhizomatic progression advocated by Deleuze and Guattari in their 1976 write-up of the rhizome concept and again in *A Thousand Plateaus*.[22] Charles describes *enracinerrance* as a "deliberately oxymoronic term" that "simultaneously takes into account rootedness and wandering, the memory of origins and the new realities of migration; it takes account of the rootedness in wandering."[23] He acknowledges awareness of Derrida's concept of *destinerrance*, which he takes to mean "a destiny of wandering" and distances himself from what he reads as

20. "C'est que la poésie de Saint-John Perse . . . augure un nouveau mode du rapport à l'Autre, qui par paradoxe, et à cause même de cette passion d'errance, prophétise la poétique de la Relation." Ibid., 54.

21. The term appears among a list of writers who influenced Charles: "l'enracinerrance de kenneth white polyculturel insatiable à voyager." Charles, *Le corps noir*, 198. Kenneth White is a Scottish-born Francophone writer who founded the *Institut international de géopoétique* in France in 1989. The forging of portmanteaus like *enracinerrance* is a signature of Charles's corpus. For example, in *Le corps noir,* Charles invents *autroubiographique* (*autobiographique* + *trou* [hole]), *Charles de gaulche* (Charles de Gaulle + *gauche* [left]), and *Charlembour* (his own name + *calembour* [pun]).

22. *Le corps noir*, whose cover calls it an essay, is a mutinous prose experiment that blends autobiography, countless citations of historical documents, pictures, pop culture references, neologisms and portmanteaux, Marxist theory, essayistic reflections, jokes, and free association. The energy, scope, and structure of this book make it comparable to *A Thousand Plateaus.*

23. Charles, "L'enracinerrance," 38.

Derrida's autobiographical appropriation of the word.[24] Charles goes on to explain that he sees himself as an *enracinerrant*, and that no other terms—he cites "migrant writer," "cosmopolitan writer," "citizen of the world," "writer without borders," "transnational writer"—could accommodate him. His regular itinerary—a large geography of perpetual circulation in the shape of a triangle whose downward tip is Haiti and whose upended base angles are the United States and France—is a tree whose roots are everywhere. Refusing the notion of an "authentic Haiti," delivered via the "constant blahblah" of those he calls the "identity cops" (39–40), Charles prefers an open-ended and multiplicitous take on what could be the character that geographic spaces lend to the people who inhabit them, an experientially additive rather than subtractive approach to travel and residence. Open-endedness and a logic of accumulation guide the spirit of *enracinerrance*, whose principle is this: "Let the entire world circulate" (39). We can read Charles's theorization of his neologism as the attempt to reconcile two historically antithetical tendencies—one to move and the other to keep still—that characterize his own identity. Drawing from experiential observations and profoundly felt pain, Charles is an authority on nomadic living. One gets the sense that his theorizations are not theoretical at all but are instead the living, empirical account of a life lived in transit.

Vilém Flusser (1920–1991), a Czech-born Jewish writer who lived for long periods in Brazil, France, Germany, and other places, pursued the theme of the creative benefits of exile, nomadism, and homelessness throughout his writing life. In his youth, he fled from Prague after the Nazi occupation in 1939. He lost most of his family and everyone in his entourage to the camps. This scholar, whose perpetual circulation began under these traumatic circumstances, found a way to reconceptualize nomadism not as a misfortune but as a gift. He argues that we should not pity "the uncounted millions of emigrants (whether [they] are guest workers, expellees, or intellectuals traveling from one seminar to another)," including the "Vietnamese in California, the Turks in Germany, the Palestinians in the Persian Gulf, and the Russian scientists at Harvard"; instead they should be considered "pioneers of the future" and "role models whose examples we follow in case we are sufficiently daring."[25] Echoing Levinas's critique of Heidegger, whose desire for a return to the root he considered residually pagan and childish, Flusser claims that the attachment to home is

24. Ibid. Later he adds, "Let's return to Derrida's proposition: he thus states it for himself and, of course, he is the only one authorized to articulate something of the deep feeling of his destiny as a wanderer" (40).

25. Flusser, "Taking Up Residence," 92.

the result of a subconscious fetal attachment to the motherland. Like Levinas, Flusser is suspicious of the secret. Concepts that must hide themselves in order to remain functional do their work surreptitiously because a full exposure would reveal the inherent weakness of the concept. Mystification is the only thing that enables their existence. Flusser, like Levinas before him, wants to shine a bright light on these weaknesses. Using metaphors of vision, Flusser describes the process of self-enlightenment with regard to the root. Blindness toward the home occurs because of the numbing effects of habit, repetition, and the banal aspects of daily living. He claims that "every home is a sacralization of the banal. . . . Whatever its shape, home is nothing more than a place to live surrounded by mysteries" (99). Home, then, is "a network of habits" (100). Flusser makes the distinction between "home" and "a home," claiming that everyone, even the homeless who live under bridges, has a home. When one is away from home, data transforms from meaningful information into meaningless noise. In an aesthetic excursus, Flusser reformulates his notion—reiterated throughout his corpus—that the cycle of aesthetics begins with the ugly (what is unfamiliar), then becomes beautiful (when the unfamiliar begins to yield information to the person who beholds it), then pretty (when beauty becomes dull), and finally ugly again (when the pretty becomes superfluous). He argues that "patriotism is a symptom of an aesthetic disease" (101), because it involves the confusion of prettiness for beauty, the belief that your home is more beautiful than anyone else's. The stranger or foreigner embodies the possibility that one's own home is just one among many, rendering the home unmagical, unspecial, and unsingular. This is, in part, why we fear the stranger. In his essay "Exile and Creativity," Flusser celebrates exile as the possibility of removing the blanket called habit, which covers everything in one's home. This unveiling precipitates a new visibility, a new possibility to glean information from one's own home. In his view, this is what connects "the Christian story of man's expulsion from Paradise and his entrance into the world, the Jewish mystic's story of the exile of divine spirit in the world, and the existentialist story of man as a stranger in the world": Each created the conditions of a creative and defamiliarized engagement with an alien world. The "ocean of chaotic information" involved in the experience of exile creates the potential for creation, which he defines simply as "data transformation."[26] Flusser asks a pointed question: "Even without intending to do so, have the expellers not done the expelled a service?" Flusser celebrates all that *un-settles*. Despite the belief on the part of rooted peoples that they make history, Flusser argues instead

26. Flusser, "Exile and Creativity," 104.

that the "uprooted make history" (105), an argument that annuls Deleuze and Guattari's Nomadology-History dichotomy. "The Jews are not part of Nazi history," he claims. "The Nazis are part of Jewish history" (107). He concludes, "Exile, no matter what form it takes, is a breeding ground for creative activity, for the new" (109). There is undoubtedly a therapeutic element in Flusser's negotiation of his own nomadism, which he probably would not have chosen under less devastating circumstances. Part of the healing process involves *owning* what was thrust upon him. Embracing his self-expulsion is part of a quiet retribution in response to Nazi brutality. He attempts to change the narrative of who actually won, showing that exile keeps the spirit agile and clear-eyed, while sedentary life leads to blindness and cultural lethargy. The nomad takes his home with him.

Because Weil and Heidegger were writing in a context long before syntheses like *destinerrance*, *errance enracinée*, *enracinerrance*, or taking up residence in homelessness had been proposed, they privileged the language of roots and were complicit in the more disquieting aspects of various pro-roots philosophies. It became clear that the root-centered debates of the period weren't just about regionalism but were rather about the importance of ostracizing the uprooted and uprooting force represented, for example, by the Jews in the minds of the nationalists. Because, as Hannah Arendt has argued, much of modern anti-Semitism is caused by the "inability to understand historical contexts,"[27] one sees in the provincialism of Barrès and his proponents a refusal to think in terms of causes rather than simplified and contextless effects. As argued in chapter 1, the lack of context—be it historical, geographic, natural, social—is the central cause of the most devastating events in the modern age. One could say that the nationalist and anti-Semitic arguments that circulated in Weil and Heidegger's time were uprooted from history, ignoring the long chronicle of Europe's relation to Judaism, urbanization, and politically motivated

27. Nor can modern anti-semitism exist without appealing to fears whose origins lie in an inability to understand historical contexts—such as the ancient, deeply rooted dread of the Wandering Jew (and the more primitive—that is, the more unversed in history the level of society is—the deeper the roots), of Ahasuerus, of these "lateborn children of death" (Clemens von Brentano), of the incomprehensible phenomenon of an ancient people who have survived so many European catastrophes without land or soil, that is, apparently without any earthly ties, who do not live in an earthly fashion and who cannot die as other peoples do, who like a ghost have rescued themselves out of times long past, in order to feed themselves like vampires upon the blood of the living.

Arendt, *Jewish Writings*, 65

circulation. Paradoxically, rootedness to an idealized past often involves a total ignorance of history. Ultimately, Barrès expresses the same paganistic separation from historical time as do Heidegger and the National Socialists. This militant paganism, what Ernst Bloch has called *pastorale militans*[28]—visible, for example, in the botanical language of Hitler's *Mein Kampf*[29]—tries to set up a configuration in which the neo-pagans have nature on their side against the enemy. Aligned with this omnipresent and indomitable force, those who worship nature celebrate a distinct advantage over those they see as alienated from nature. As we will see, Blanchot and Levinas noted this paganistic tendency in right-wing politics and proposed a celebration of Judaism's nomadism and its tendency away from nature worship. The long history of the *völkische Bewegung*—the populist movement in Germany that conjugated the natural world, folklore and mysticism, romantic nationalism, and bloodlines—constitutes what Blanchot and Levinas see as the actual uprooting force in Europe. It is not, in their view, the Jews who have uprooted Europe. On the contrary, it is those who have invested the land with a holiness that causes alienation, turf wars, and other forms of primitivistic exclusion. This *Blut und Boden* mythology does not belong exclusively to Germany; we find it even in

28. Ernst Bloch writes, "Whereas from the country old sap rises into long-forgotten shoots, it nourishes National Socialists and folkish mythologists, in short, arises as *pastorale militans*"; and "Rootedness in the soil which has become mythical thus not only produces false consciousness but strengthens it through the subconscious, through the really dark stream." Bloch, *Heritage,* 48, 51.

29. Hitler writes, "If a people refuses to guard and uphold the qualities with which it has been endowed by Nature and which have their roots in the racial blood, then such a people has no right to complain over the loss of its earthly existence" (Hitler, *Mein Kampf*, 282); "[The Nazi Movement] was not meant to be an organization for contented and satisfied people, but was meant to gather in all those who were suffering from profound anxiety and could find no peace, those who were unhappy and discontented. It was not meant to float on the surface of the nation but rather to push its roots deep among the masses" (285); "When the individual is no longer burdened with his own consciousness of blame in this regard, then and only then will he have that inner tranquility and outer force to cut off drastically and ruthlessly all the parasite growth and root out the weeds" (36); "The man who thinks that he can bind himself by *treaty with parasites* is like the *tree* that believes it can form a profitable bargain with the ivy that surrounds it" (557); "All those movements which owe their expansion to a so-called combination of similar organisms, which means that their external strength is due to a policy of compromise, are like plants whose growth is forced in a hothouse. They shoot up externally but they lack that inner strength which enables the natural plant to grow into a tree that will withstand the storm of the centuries" (303).

the *Marseillaise*, the French national anthem, albeit as an evocation of the enemy's spilt blood and not that of the French themselves:

> Marchons, marchons,
> Qu'un sang impur
> abreuve nos sillons
> (Let us march, let us march,
> Til their impure blood
> Soaks our furrows)

Yet the adjective "impure," used to describe foreign blood, participates in the same logic as discourses of Aryan supremacy, eugenics, and other National Socialist tropes of racial primacy. The most evil acts in twentieth-century Europe were precipitated by root thinking. What exactly grows from fields soaked with enemy blood?

The relationship of Weil and Heidegger to Judaism is crucial to a full account of how and why these thinkers both came to embrace a vocabulary of roots in the period surrounding the war. Numerous fragments from Weil's writings make it clear that she viewed Jews as a rootless people and that she had a distinct distaste for the violence in the Hebrew Bible; she blamed Jews for everything from the Inquisition to twentieth-century capitalism and totalitarianism.[30] Weil's position that they were responsible for uprooting Western

30. Here are a few examples: "Religious thought is genuine whenever it is universal in its appeal. (Such is not the case with Judaism, which is linked to a racial conception)" (Weil, *Need for Roots*, 93); "The Jews—that handful of uprooted individuals—have been responsible for the uprooting of the whole terrestrial globe. The part they played in Christianity turned Christendom into something uprooted with respect to its own past. . . . Capitalism and totalitarianism form part of this progressive development of uprooting; the Jew-haters, of course, spread the Jewish influence. The Jews are the poison of uprooting personified. But before they began uprooting by spreading this poison, Assyria in the East and Rome in the West had already started doing so by the sword" (Weil, *Notebooks*, 2:575–76);

> The Hebrews, having rejected the Egyptian revelation, got the God they deserved—a carnal and collective God who never spoke to anyone's soul, up to the time of exile. . . . Is it surprising that there should be so much evil in a civilization—our own—which is corrupted at its roots, in its very inspiration, by this atrocious lie? The curse of Israel weighs upon Christendom. The atrocities, the exterminations of heretics and of unbelievers—all this was Israel. Capitalism was Israel—(and is so still, up to a certain point . . .). Totalitarianism is Israel (more particularly so among the latter's worst enemies). (570–71)

civilization stands in opposition to Levinas's argument that Christianity uproots the individual.[31] She was intimately familiar with the term *déraciné* and perhaps subconsciously took this term as an insult to her own person, given her Jewish lineage. If Weil was really a self-hating Jew, as she has been dubbed by so many critics, it is reasonable to suggest that she valued the very quality her ancestors could never achieve in the minds of many Europeans, namely, rootedness. And even if, as Levinas and Palle Yourgrau have argued, Weil was not an anti-Semite but someone who disliked the brutality of certain episodes of the Hebrew Bible and who opposed in any form the election or supremacy of one race over another, the term *déraciné* as an adjective to describe the Jews in France was omnipresent in the period in which she lived, from the echoes of anti-Dreyfusard rhetoric to Barrès's *Les déracinés*. The metaphor was ineluctable. In Heidegger's case, his adherence to an anti-Semitic ideology is clear. Stereotypes about Judaism led Weil and Heidegger, consciously or unconsciously, to their rhizophilic stance and their application of a pro-roots philosophy to matters that had ostensibly nothing to do with the Jews. In other words, the trope of the wandering Jew—*le Juif errant* or *der Ewige Jude*—which both thinkers undoubtedly encountered early in their studies and as a central trope in the political debates of the period, initiated their worldview that humans are rooted creatures. Without this commonplace, it is unlikely that they would have arrived nearly simultaneously at these strikingly similar motifs of rootedness and groundedness. They were not the only ones touched by the trope. The myth of rootedness was in the air; scientists, writers, philosophers, and politicians referred to it in some form, from ethnogeographic studies to regionalist novels to wartime propaganda, nearly always in reference to a politics of belonging.

For Heidegger, the Jew represented what was *unfaßbar* (intangible, ungraspable, or inconceivable), a quality he claims led to Europe's woes: "World

---

For more on this debate, see "Simone Weil against the Bible," in Levinas, *Difficult Freedom*; the chapter "On the Jewish Question" in Palle Yourgrau, *Simone Weil (*London: Reaktion, 2011); Palle Yourgrau, "Was Simone Weil a Jew?," *Partisan Review* 68, no. 4 (Fall 2001): 629–41; Thomas Nevin, *Simone Weil: Portrait of a Self-Exiled Jew* (Chapel Hill: University of North Carolina Press, 1991).

31. "This double attitude shows why revolutionary Christianity, which uprooted the individual from the strongest links binding him to his condition, was horribly conservative: bowing before the established order, afraid of scandal, paradoxically associating its horror of a nature of grace with the poetry of naivety, fields of wheat, the virtues of being warrior-like and putting down roots, of being a man-plant, a humanity-forest whose gnarled joints of root and trunk are magnified by the rugged life of a countryman." Levinas, *Difficult Freedom*, 100.

Jewry, spurred on by the emigrants who've been allowed out of Germany, is intangible everywhere and, as much as it develops its power, never has to take part in activities of war, whereas the only thing left for us is to sacrifice the best blood of the best of our own people."[32] Again, the *Luftmenschen* trope, alluded to in chapter 2, takes shape here as the picture of a people who slip through the fingers, living abstractions whose lack of concreteness still manages to hammer Europe apart. Shunning the accusation that anti-Semitism is a racial question, he transforms it into a metaphysical question, claiming that Jews are responsible for the uprooting of Being.[33]

Another root fixation that appealed to both Weil and Heidegger was an attachment to Greece as the cultural cradle of Germany and France. When seeking nostalgic inspiration from their forebears, the Europeans can always turn back to the Greeks.[34] One of the most stable aspects of Heidegger's corpus is his Graecophilia.[35] In 1924 he announced to his students: "We need to win back rootedness and autochthony as it was alive in Greek science."[36] Deleuze and Guattari claim that the confused attempt to get to the origin or the ground,

32. Heidegger, *Gesamtausgabe 96*, 262.

33. "Even the thought of an agreement with England, in the sense of a division of imperialist 'jurisdictions,' does not reach the essence of the historical process that England is now playing out to its end within Americanism and Bolshevism, and this at the same time means within world Jewry. The question of the role of world Jewry is not a racial question, but the metaphysical question about the kind of humanity that, without any restraints, can take over the uprooting of all beings from being as its 'world-historical' task." Ibid., 243.

34. Hans-Georg Gadamer—around a decade younger than Heidegger, who was his teacher—was well known for his Graecophilia. In his texts *The Beginning of Philosophy* [*Anfang der Philosophie*] and *The Beginning of Knowledge* [*Anfang des Wissens*], he emphasizes the indebtedness of modern European thought to the Greek forebears. Paul Allen Miller has argued that "the works of Lacan, Derrida, and Foucault, however revolutionary in their intent, cannot be properly understood outside the context of their own profound indebtedness to classical and particularly Hellenic culture." Miller, "Classical Roots," 204.

35. Of Graecophilia in Hegel and Heidegger, Deleuze and Guattari write, "What remains common to Heidegger and Hegel is having conceived of the relationship of Greece and philosophy as an origin and thus as the point of departure of a history internal to the West, such that *philosophy necessarily becomes indistinguishable from its own history*. However close he got to it, Heidegger betrays the movement of deterritorialization because he fixes it once and for all between being and beings, between the Greek territory and the Western earth that the Greeks would have called Being." Deleuze and Guattari, *What Is Philosophy?*, 95. They continue: "[Heidegger] wanted to rejoin the Greeks through the Germans, at the worst moment in their history: is there anything worse, said Nietzsche, than to find oneself facing a German when one was expecting a Greek?" (108–9).

36. Unpublished text quoted and translated in Bambach, *Heidegger's Roots*, 17.

to recognize in the Germans the lost greatness of the Greeks, caused Heidegger to mistakenly take "not only the German for a Greek but the fascist for a creator of existence and freedom." In short, they argue: "He got the wrong people, earth, and blood."[37] Charles Bambach outlines in detail the forms of Heidegger's attachment and indebtedness to the Greeks, using Nietzsche's philhellenism as the bridge between them. He summarizes the complex lineage of Heideggerian thought in this passage: "In the curious and fateful conjoining of Nietzschean *Kampf* and Hölderlinian *Heimat*, of Heraclitean *polemos* and Eckhartian *Gelassenheit*, there emerges a Heideggerian philosophy of roots that both brings together and sets asunder the counterposed realms of the martial and the pastoral."[38]

This summary makes immediately clear some of the salient similarities and differences between Weil and Heidegger's rhizophilia. What binds them is their mutual attraction to Greece and the belief that Europe must look back to Greece to save itself. Both Weil and Heidegger were greatly influenced by Meister Eckhart's notion of detachment (*Abgeschiedenheit*), which in Weil's work takes the form of *décréation* and in Heidegger's, *Gelassenheit*. Both thinkers were highly suspicious of the bourgeois class and both lamented the cultural legacy of the French Revolution. Both were opposed to the Treaty of Versailles, which they believed to be one of the worst tools of humiliation in the history of Europe and responsible in many respects for the gravity of World War II.[39] The Hellenist Nicole Loraux has argued that for the Greeks themselves, humans were imagined to have come directly from the earth in one of two ways, either born directly "out of the soil like a plant or a child from the womb . . . ; according to other myths . . . the human creature, made of earth and fashioned by an artisanal god, is the product of a manufacturing."[40] Using this distinction, we might say that Heidegger pursues in his writing the lineage

37. Deleuze and Guattari, *What Is Philosophy?*, 109.

38. Bambach, *Heidegger's Roots*, 9–10.

39. In a letter to Georges Bernanos, Weil writes, "I was ten when the Treaty of Versailles was signed. Until then, I'd been a patriot with the same elation as all children during wartime. The will to humiliate the defeated enemy in such a repugnant manner, which brimmed over everywhere at that time (and in the years that followed), cured me once and for all of this naïve patriotism. The humiliations inflicted by my country are more painful to me than the ones it can endure." Weil, *Ecrits*, 224.

40. "L'homme provient de la terre. . . . Mais il est pour l'humanité au moins deux façons de provenir de la terre: dans certains mythes, tel le mythe platonicien des *gêgeneis* . . . tels aussi les mythes athénien ou thébain d'autochtonie . . . l'homme . . . surgit de la terre comme une plante sort du sol ou l'enfant de la matrice; selon d'autres mythes, tel le récit hésiodique de la création

of the first, while Weil combines both of these myths in her conception of human provenance. Her Christian faith nurtures the idea of a sculptor-God; her fondness for early Greek culture and its foundational myths fuses the idea of humanity as an earthen race, either as autochthonous creation or as sculpture.

WEIL'S FEAR OF ABSTRACTION

> One must uproot oneself; cut down the tree and make of it a cross, and then carry it always.
>
> SIMONE WEIL, *Notebooks*

Simone Weil's legacy is an unstable one. She has been depicted as all of the following: Christian mystic, committed Marxist, Platonist, factory worker, intellectual genius, self-hating Jew, androgyne, unintentional feminist, anorexic, and madwoman. Much of the academic interest in her work arose posthumously with the publication of her unfinished reflections, such as *La pesanteur et la grâce* (*Gravity and Grace*), *L'enracinement* (*The Need for Roots*), and *Attente de Dieu* (*Waiting on God*). Her thought has been particularly well received in Italy, perhaps because of the compatibility between a certain kind of Italian Catholic thought and the specific orientation of her Christian mysticism. Many of her contemporaries found her to be an eccentric (in both the negative and positive connotations of the term) and an original thinker with an unconventional, highly self-invested approach to politics and religion. Her mythology has continued to grow, and her work today is appreciated by contemporary European and American audiences, particularly by those with an interest in feminism, Catholicism, and labor movements. Her totalizing diagnostic practice looked everywhere in culture past and present for the causes of the most pressing problems in her time. Perhaps the most fascinating aspect of her biography is her refusal to describe culture from afar. She chose instead to embed herself in unfamiliar experiences. For example, once she took an interest in the daily experience of workers, she wasn't content to draw conclusions from written testimonies or descriptions of factory life. She went to work in a Renault car factory herself. She also worked alongside the peasants in the vineyards in Saint-Marcel-d'Ardèche in southeastern France. She did not exempt herself from physical labor simply because she was a trained intellectual. The depth and breadth of her knowledge and the variety of experiences she'd

---

de la femme, la créature humaine . . . faite de terre et modelée par un dieu artisan, est le produit d'une fabrication. Terre-champ, terre-matière: le glèbe ou la glaise." Loraux, *Né de la terre*, 9–10.

acquired by the time of her death at age thirty-four are immense by today's standards. She was fluent in Greek by age twelve and later learned Sanskrit and studied religions such as Hinduism and Buddhism. She had a panoramic understanding of European and world history and was accomplished in mathematics and philosophy. She briefly joined a French-speaking anarchist militia on the Republican side in the Spanish Civil War. Leafing through Weil's notebooks, one finds mathematical equations with prayers and reflections on God in between, connections between lived experience and intellectual life, etymological analyses, quotations in Greek, aphorisms, excerpts from poems, lists of philosophical questions, and reflections on almost every discipline (biology, physics, mathematics, political theory, etc.). Richard H. Bell describes Weil's corpus this way: "The 'primeval chaos' of Simone Weil's world is the world of Greek poetry and tragedy and Greek philosophy; it is the world of St. John of the Cross and Meister Eckhart; it is the world of a radically incarnational Christology without dogma; it is a world of Descartes, Kant, Racine, Alain, and René Le Senne; it is the world of the *Bhagavad-Gita*, Taoism, and Zen Buddhism; it is a world of folklore and mythology."[41] Perhaps contemporary readers are drawn to this variety of sources and experiences that informed her work, coupled with the absolute urgency of voice with which she expresses her religious, philosophical, and political ideas. This same urgency is repellent to other readers who hear in her voice evidence of madness and an unyielding, perhaps autocratic character. Weil's emotional disposition was perceived as being at odds with the intellect and was thus unappealing to many among her scholarly contemporaries.

I want to make the claim that Weil's fear of rootlessness can be attributed to her general fear of abstraction. Rootedness for her meant to turn one's full attention toward a concrete, spiritually infused world, participating actively to sustain the past, present, and future of a community, and to nourish one's environment in order to strengthen the moral, intellectual, and spiritual aspects of life. Conversely, to be rootless is to face a set of circumstances in which things are replaced by figures of those things. The world was becoming rootless owing to the simple fact that "the relation of the sign to the thing signified [was] being destroyed."[42] Unlike the extraction-fearing majority of thinkers presented so far in this study, who fear being plucked from their contexts, Weil's most explicit fear was abstraction, illustrated in her antipathy toward money, algebra, *machinisme*, and the conversion of labor into abstract value.

41. Bell, introduction to *Weil's Philosophy*, 21.

42. Weil, *Gravity and Grace*, 209.

Despite his immateriality, God was no abstraction for Weil; she sensed his presence bodily and considered the earth a manifestation of his grace. She judged the Enlightenment as a progression toward cold and hard reason, so far removed from the warmth of Christian love that she lauded time and again in her writings. It is fair to say that Weil anticipated Gianni Vattimo's notion of weak thought (*il pensiero debole*), not simply as an abstract concept but as her lived approach to the world through God.[43] As an intellectual, she felt ill at ease remaining in the walled-off world of academic political and economic theory; her dread of abstraction sent her into the factories and fields to work alongside the laborers who remained alien to many intellectual Marxists. As she read the news of the war from England, she prevented it from becoming an abstraction by constraining herself to the same rations as the French soldiers. *The Need for Roots* is a particularly strong example of this antipathy toward abstraction, as it shows a thinker who wants to help her compatriots undo the alienation of the modern age by guiding them toward a kind of self-grafting into vocation, education, faith, country, and the land. It is too reductive to say that Weil simply feared modernity as such. It was rather the tendency of capitalism, intellectualism, and bureaucracy to multiply abstractions that disquieted her. I imagine that cyberwarfare and electronic trading would have terrified Weil much more than a drone.

Weil can be described as a vertical, symmetrical thinker. Aside from her attraction to the vertical sacred of Christianity, she believed heavily in the necessity of symbolic hierarchies as a vital need of the soul. In her view, the natural order of things involves a vertically polar structure of superior and inferior. However, she also considers equality a necessity of the human soul, arguing that the worker is not less than his supervisor, simply different from him. Both are necessary. She claims that in exchange for the privilege of supervising others, the supervisor takes a greater risk in maintaining this position and is more accountable if his workers have grievances against him. Conversely, the worker, whose vocational risk is inherently smaller, should suffer less when his supervisor has grievances against him. The entirety of Weil's thought advocates the maintenance or recovery of equilibrium. For every right, there is a

43. Weak thought can be understood as the antithesis of the strong thought of metaphysics, which seeks—self-delusionally, in Vattimo's view—to discover or confirm unshakable truths. It is worth noting that Vattimo's development of *il pensiero debole* overlaps with his attempts in the 1980s to reconcile Christianity and postmodernity. For more on Weil and weak thought, see Alessandro Dal Lago, "On the Ethics of Weakness: Simone Weil and Nihilism," in *Weak Thought*, ed. Gianni Vattimo and Pier Aldo Rovatti, trans. Peter Carravetta (Albany, NY: SUNY Press, 2012), 111–37.

responsibility. For every sin, there is a corresponding punishment. The root, then, as the mirror image of the branches, fits neatly into her symmetrical and vertical notion of justice. Unlike Gide's immanent *nourritures terrestres*, Weil's *nourritures* are purely transcendental.

In the early 1940s, Weil, who was working as an editor at France Libre in London, drafted an ethical guide for the French people, a document that would reorient the nation after the failures leading up to the war. The resulting treatise was never completed but was published posthumously under this title given by the editors: *L'enracinement: Prélude à une déclaration des devoirs envers l'être humain* (*The Need for Roots: Prelude to a Declaration of Duties toward Mankind*). The text elicited highly varied reactions, including praise, ambivalence, and scorn. Albert Camus considered it one of the most important books to have appeared after the war.[44] In a review, Georges Bataille noted that the ideas put forth in *The Need for Roots* would be virtually impossible to execute,[45] but he conceded that the zeal and passion of Weil's work was compelling enough to captivate the reader.[46] Maurice Blanchot, in his essay "Affirmation (desire, affliction)," paints an ambivalent portrait of the contradictions

44. Camus continues, "In any case, it seems impossible to me to imagine a rebirth for Europe that does not take into account the demands defined by Simone Weil in *The Need for Roots.*" Camus, *Œuvres*, 391–92. John M. Dunaway argues that nearly all of Camus's characters are *déracinés* in Weilian terms. See John M. Dunaway, "Estrangement and the Need for Roots: Prophetic Visions of the Human Condition in Albert Camus and Simone Weil," *Religion and Literature* 17, no. 2 (Summer 1985): 35–42. Fred Rosen, too, has noted Weil's lasting influence on Camus. See Fred Rosen, "Marxism, Mysticism, and Liberty: The Influence of Simone Weil on Albert Camus," *Political Theory* 7, no. 3 (August 1979): 301–19.

45. "Furthermore, we could say that if it were necessary to judge *The Need for Roots* from a practical angle, this book would not hold one's attention for long. Many formulations have in sight the narrow interest I spoke of: these are hackneyed, inapplicable, or paradoxical slogans. If the vigor of the expression obscures its weakness, it can do so only for instant." Bataille, "La victoire militaire," 793.

46. "It is not Simone Weil's intelligence but her emphasis that pulls one in: it is through an extraordinary tension that her book strikes with such force. This excessive zeal and this default to the idea of failure [*manquement*] are signs of a passion analogous to the one a lover feels for the loved one, so that the fear of loss is so great as to wrench the ground." Ibid., 792. Here, I have translated "manquement" with the admittedly vague "failure," as the notion of *manquement* in Weil's work is ambiguous. Bernard Halda asks, "What distinction does Simone Weil make between error, breach of an intellectual rule, fault, violation of a moral rule, sin, and violation of the laws to specific to human destiny, it being evident that in Christianity, the law remains in accordance with the moral order?" Halda, *L'évolution*, 142.

and stubborn certitude in Weil's thought.[47] Conor Cruise O'Brien, an Irish politician, wrote of *The Need for Roots* in the early 1980s: "The atmosphere she evokes is that of a state governed by a spiritual and moral elite, a rule of the saints. . . . This is the rather discouraging outcome of a hypothetical effort to apply in politics the thinking of a writer who was essentially nonpolitical, and even antipolitical."[48] His reaction sums up the paradox of Weil's late work, namely that she continued to write heavily political texts during perhaps the most politically agitated moment in twentieth-century European history, even after her purported renunciation of politics in favor of spiritual thought. For this reason, her solutions to Europe's problems are of a quasi-metaphysical nature and thus nearly impossible to implement in any practical way. The spirit of these texts, however, was meant to galvanize Europeans to abandon their course for a new one, perhaps most strongly through *pathos*. *The Need for Roots* could be described, then, as an inspirational text, but not a pragmatically executable one. Unlike Denise Levertov's skepticism seen in chapter 2 regarding the human's rerooting of itself in the environment, Weil's utopianism leads her to believe that, with some spiritual guidance, Europe could be restored. She set out to provide this guidance herself. But in this respect, Weil encounters the same problem as Marxism in general, that is, the difficulty of translating theory into action.

Her sweeping diagnosis in *The Need for Roots* attributes nearly every societal predicament to what she calls *déracinement*, or uprootedness. The root metaphor merges verticality, symmetricality, and the nutritive aspect of the sacred, a key component of Weil's Christian poetics. Many of the fragments in her notebooks focus on the conjugation of the botanical and the human.[49] Her botanical sympathies lead unsurprisingly to the conclusion that a new rootedness is the only way for the European to reintegrate into God's world. The need to address a loss of roots in Europe derives from her belief that the condition is pathological and transmissible, capable of infecting every person

47. See Maurice Blanchot, "L'affirmation (le désir, le malheur)," in *L'entretien infini* (Paris: Gallimard, 1969).

48. O'Brien, "Patriotism," 95–96.

49. For example, she writes, "A man does not perform the same actions if he gives his consent to obedience as if he does not; just as a plant, where everything else is equal, does not grow in the same way if it is in light as if it is in the dark. The plant does not have any control or choice in the matter of its own growth. As for us, we are like plants which have the choice of being in or out of the light." Weil, *Waiting for God*, 129–30. She also makes regular use of Aristotle's notion of the vegetative soul, the only common soul shared by plants, animals, and humans, which is responsible for reproduction and growth.

and every aspect of life. She establishes this general principle: "Whoever is uprooted himself uproots others. Whoever is rooted himself doesn't uproot others."[50] A common trope in Weil's corpus is that of bridging or approximating, in the sense of bringing closer together. As David McLellan writes, "*The Need for Roots* is about the preservation of the *metaxus* as bridges leading to the supernatural."[51] Abstraction resists the bridging reflex since it involves a drawing away from (*ab* + *trahere*); thus it is a phenomenon of distancing. She pictured the world becoming ever more remote from itself, replacing the real with empty figures. To restore the broken bonds between the Chthonic and the Uranic, between people and their livelihoods, she promotes the spiritualization of work, the regrounding of education, and the ownership of property, thus clearing away the distracting mediations that hold the person at a distance from God's world. In her conception, the severing of roots is a figure for the unchecked multiplication of intervening distractions from truth.

Ultimately, her key recommendation for reestablishing them relies on the imperative that these broken synapses in the human soul must be rewired through physical labor and its spiritual benefits. Weil's obvious pedagogical intent and her professed belief in certain principles of pedagogy (the importance of repetition, the meaningful association of physical gestures and corresponding thoughts or words) manifest themselves particularly in the passages on labor and education. Suffering from a general feeling of uprootedness not only while in the factory but also at home, in political parties, in unions, in places of leisure, and in an intellectual milieu, the alienated worker has come to feel ashamed of his vocation. As a solution, Weil proposes a progressive spiritualization of bodily gestures, particularly those involved in one's vocation. Ann Pirruccello calls these Weil's "somatic practices" and notes their proximity to some aspects of Buddhist meditation.[52] Weil evokes the image of a peasant farmer who, as he tends his fields, preserves in the back of his mind the biblical references to the seed in order to associate a spiritual reflex with the physical gestures of his labor. One's vocation can be schematized into a series of repetitive motions, particularly in the case of factory workers or farmers, who must perform the same motions again and again to survive. She explains that the youth aiming to become laborers in factories or other urban settings

50. Weil, *Need for Roots*, 48.

51. McLellan, *Utopian Pessimist*, 258. For Weil, "the *metaxus* were links with God, bridges toward the transcendent" (201).

52. See Ann Pirruccello, "Making the World My Body: Simone Weil and Somatic Practice," *Philosophy East and West* 52, no. 2 (October 2002): 479–97.

could use mechanics as their organizing analogy, while youth aiming for farming or other rural labor jobs could be trained primarily through analogies of the cycles of the natural world. She literalizes the notion of rootedness by advocating property ownership for both urban workers and rural peasants. In her view, a sense of rootedness can be had only through the possession of land, through its tending and care, and through the pleasure procured when the fruits of one's labor have been harvested. By extending and fusing the self with one's material environment, the human imitates the root's purposeful possession of its environment.

Regarding education, Weil claims that young people come of age as cultural orphans with no stable heritage to ground them. They lack a comprehensive education that emphasizes the cultural contributions of the Greeks. The education system fails students in another way: it draws no connections between taught material and the real world. Geometry, for example, is presented as an abstract discipline with no consequences in everyday life. She points to the absurdity of an educational system that privileges rote memorization over contextualization, in which children can brainlessly quote astronomical theories but never bother to look up at the sky.[53] Knowledge acquired through formalized schooling has become an abstract exercise devoid of material relevance. She also embeds a critique of the colonial endeavor in her grievances against French education, namely that schooling should address the specificities of a given culture. For the French to indoctrinate the children in their colonies with a mistaken cultural heritage, intentionally or as mere oversight, underscores the nation's ignorance about the universal human need for cultural rootedness. Such ignorance makes it impossible for anyone living under France's colonial rule to develop a sense of cultural belonging, and the result has nefarious consequences for the *métropole*. Weil expands her educational critique to include the problem of class, claiming that anyone not born into a middle-class, intellectual milieu is automatically excluded from any kind of intellectual engagement. Peasants and workers are alienated by the exclusive vocabulary and uninviting atmosphere of higher education. By failing to "translate" into

53. A lot of people think that a little peasant boy of the present day who goes to primary school knows more than Pythagoras did, simply because he can repeat parrot-wise that the earth moves round the sun. In actual fact, he no longer looks up at the heavens. This sun about which they talk to him in class hasn't, for him, the slightest connexion with the one he can see. He is severed from the universe around him, just as little Polynesians are severed from their past by being forced to repeat: "Our ancestors, the Gauls, had fair hair."

Weil, *Need for Roots*, 46

a language that is comprehensible to workers and peasants the important concepts from philosophy, literature, and other disciplines, the intellectual class has precluded the civic engagement of a sizable proportion of the total population. A celebration of local folklore has the potential to "help the peasants feel at home again in human thought" (87).[54]

In her assault on abstraction, she pursues a very old idea, found already in the New Testament, in the first epistle of Saint Paul to Timothy: *Radix omnium malorum cupiditas*, "For the love of money is a root of all kinds of evil." She blames it in part for both world wars and for the progressive root loss of the European people. In *Gravity and Grace* (*La pesanteur et la grâce*, written in the early 1940s and published posthumously in 1947), she writes that money, alongside algebra and *machinisme* (the mechanization of the means of production), is one of the three forces responsible for civilization's unraveling. In this aspect of her social criticism, we see a clear link between Christianity and Marxism, both wary of money as a force of moral corruption and as a tool of control. For very different reasons and with very different strategies, these two utopian ideologies urge their adherents to deal cautiously with the human-made danger of money. It is clear that Weil deploys both Christian and Marxist rhetoric to make her case against coveting it.

What I believe to be Marx's most interesting statements about money, and, probably not coincidentally, the most literary, are found in an early manuscript called "The Power of Money in Bourgeois Society," from 1844. In this text, Marx invokes Goethe's *Faust* and Shakespeare's *Timon of Athens* to argue that money has a magical power of reversing nature, allowing the rich man to make his what he does not naturally possess. Goethe's Mephistopheles says:

> Six stallions, say, I can afford,
> Is not their strength my property?
> I tear along, a sporting lord,
> As if their legs belonged to me.[55]

54. She is more specific about exactly what this social class should be learning: "[Peasant students] should be told about the part played by shepherds in the first speculations made by the human mind, those concerning the stars, and also—as the comparisons which continually occur in ancient texts indicate—those concerning good and evil. They should be made to read peasant literature—Hesiod, *Piers Plowman*, the complaints of the Middle Ages, the few contemporary works which are of authentically peasant inspiration; all this, of course, without forgetting the claims of general culture." Ibid., 88.

55. Goethe quoted in Marx, "Power," 166.

He then quotes Shakespeare's Timon, who claims that "yellow, glittering, precious" gold has the power to "make black white, foul fair, wrong right, base noble, old young, coward valiant."[56] From these two literary examples, Marx goes on to expound on his claim that money has the power to reverse nature. He writes: "I *am* ugly, but I can buy for myself the most *beautiful* of women. Therefore I am not *ugly*, for the effect of *ugliness*—its deterrent power—is nullified by money. I, as an individual, am *lame*, but money furnishes me with twenty-four feet. Therefore I am not lame. I am bad, dishonest, unscrupulous, stupid; but money is honored, and hence its possessor." He proceeds to call money "the universal *agent of separation* as well as the *true binding agent*," because everyone is subject to its network, but its influence is one of estrangement and rupture. Money distorts, confounds, and overturns "all human and natural qualities" (167). It converts impossibilities into possibilities. Here, he adds to the equation the issue of imagination; money essentially converts the "*image into reality* and *reality into a mere image*" (168). Money is creative in the sense that it forges new realities that could not exist without it, but the foundations of these new realities are purely fictional. Weil's critique of money focused on its status as a product of the imagination. She shares with Marx a semiotic understanding of money as an inverter of signs. Her claim that "the relation of the sign to the thing signified is being destroyed" by money is consistent with Marx's distress over the wealthy bourgeois man who could buy all of the qualities he didn't possess naturally, reversing nature and thus corrupting an entire system of signs.[57] In Weil's framework, "the game of exchanges between signs is being multiplied of itself and for itself. And the increasing complication demands that there should be signs for signs" (139). The world becomes profoundly less readable under such conditions.

It may seem clear why she would find a problem with money and mechanization, but why would she be opposed to algebra? What bothers her is the very difference that separates algebra from arithmetic, namely that algebra relies on abstractions, such as letters that stand in for certain variable values. She preferred geometry to algebra, because its forms had an obvious connection to the concrete world. Algebra separates the human from reality by turning thought into a series of abstract operations with no immediately visible bearing on the physical world. She claims that both algebra and money are "levellers" and that "vertical distances are not represented in them."[58] Given her obsession

56. Shakespeare quoted in ibid.
57. Weil, *Gravity and Grace*, 209.
58. Weil, *Notebooks*, 144.

with the vertical sacred, this leveling force that makes false equivalencies and corrupts the relation between signified and signifier could do nothing but destroy roots.

Weil dedicates a portion of *The Need for Roots* to the national specificities that led to Europe's crisis. She describes the easy capitulation of France to Germany as a passive symptom of uprootedness, while Germany's own violent conduct was an aggressive version of the same affliction.[59] Weil goes on to criticize France for its institutionalization of prostitution, its shameful treatment of foreigners, and its colonial greediness. She argues that France may have to choose "between her attachment to her Empire and the need to have a soul of her own again" (167). One of her more scathing criticisms of France involves its nostalgic attachment to the Roman heritage of conquest and "grandeur in the style of Corneille" (175). This celebration of the wrong values, which closely resembles Germany's cult of the hero, is largely responsible for France's attempts to expand its territories and propagate its global cultural influence. She fears that the similarities between German hero worship and French imperial fetishism point toward France's imminent mimicry of the National Socialist power play, already discernible in the Vichy Regime and its self-aggrandizing rhetoric. In support of a new kind of patriotism, she argues that the suitable attitude toward one's country is not one of heroic grandeur but of simple compassion or pity. Looking toward Christ and Joan of Arc as models,[60] Weil argues that Christian humility helps citizens behave toward their nation with the same care, tenderness, and warmth as they would toward their own children or toward their aging parents. In opposition to this kind of love, she saw the Enlightenment value of "liberty without supernatural love" as "something purely empty, a mere abstraction, without the slightest possibility of ever becoming real."[61] In general, she saw the French as dupes and claimed that "the love Frenchmen have for abstract logic makes it very easy for them to be deceived

59. A tree whose roots are almost entirely eaten away falls at the first blow. If France offered a spectacle more painful than that of any other European country, it is because modern civilization with all its toxins was in a more advanced stage there than elsewhere, with the exception of Germany. But in Germany, uprootedness had taken the aggressive form, whereas in France it was characterized by inertia and stupor. . . . On the other hand, the country which in the face of the first wave of German terror behaved far and away the best, was the one where tradition is strongest and most carefully nurtured, that is to say, England.

Weil, *Need for Roots*, 49

60. The revived Front National, led by Marine Le Pen, has begun to use Joan of Arc as a rallying figure in its publicity campaigns.

61. Weil, *Notebooks*, 2:466.

by labels."[62] She disapproved of the Encyclopédistes' elusive quest for progress, which had filtered into contemporary France and "killed any chance of inspiration being sought in a revolutionary tradition." She considered the lot of them to be "rootless intellectuals" (110). Barrès, in his *Les déracinés*, also describes the fear of intellectualism as a fear of abstraction.[63] Abstract reason divorces people from the material world and leads them into a relativizing logic through which conclusions can be made without reference to the concrete world. Yet, while Barrès binds this fear of abstraction to a larger nationalist argument, Weil is much less concerned with administrative borders or even cultural specificity. She is interested in universal *enracinement*, caring more about the fate of the world than about France's destiny.

We find in Weil's metaphorization of the root an anxiety about the transformation of her culture into a culture of abstraction. This nuance is essential and distinguishes Weil from most of the authors presented thus far. Her root-seeking is not nationalist, regionalist, or environmental. Her arguments do not correspond cleanly to the political and ecological discourses of earlier chapters. Even her Christianity is of a very different nature than Claudel's, for example. Weil was a walking exception to almost every rule imaginable. Her root metaphor, too, is exceptional. Through it, one is able to think of the world along a different spectrum, with the concrete at one end and the abstract at the other. Undeniably, the world relies far more on abstractions now than even in Weil's time. The general sense of alienation that accompanies digital living, capitalist abstraction, and the market logic of the new university might be most aligned with Weil's sense of alienation, definitely more than with Heidegger's. A decade after the war's end, he proposed a solution to another kind of European crisis through the metaphor of groundedness (*Bodenständigkeit*). While the philosophies of Weil and Heidegger have rarely been put in conversation, they contain significant, overlapping features relevant to the problem of European rootedness in the first half of the twentieth century.[64] Still, in this juxtaposition, we come to understand that even while seemingly in dialogue, they are having very different conversations.

62. Weil, *Need for Roots*, 119.

63. "Déraciner ces enfants, les détacher du sol et du groupe social où tout les relie pour les placer hors de leurs préjugés dans la raison abstraite, comment cela le gênerait-il, lui qui n'a pas de sol, ni de société, ni, pense-t-il, de préjugés?" Barrès, *Déracinés*, 21.

64. Among the few critics who have put Weil and Heidegger's thought in dialogue are Maria Villela-Petit, "Simone Weil, Martin Heidegger et la Grèce," *Cahiers Simone Weil* 26, no. 2 (2003): 181–218; and Henry Le Roy French in his chapter "Heidegger, Science, and Technology," in *Simone Weil and the Intellect of Grace* (New York: Continuum, 1999).

HEIDEGGER THE *TERROIRISTE*

> . . . sounds
> so close to terror you'd confuse
> the two, as if the finest and
> the rarest blend would come with just
> a hint of fear or pain, the sting
> and shiver of revulsion with
> the savor of the earth and sun,
> of this once, not returning, sung
> for this one ear, on this one tongue.
>
> ROBERT MORGAN, *Terroir*

Martin Heidegger was a *terroiriste*. His philosophical bioregionalism closely resembles the notion of *terroir*, a specifically French term that has no translation in English, German, Spanish, Italian, or many other languages. Thomas Parker defines terroir as "the soil, climate, and physiographic composition of a certain locale. It is a singularly French notion explaining how geographic origin influences its agricultural produce, imbuing food and wines with specific flavors."[65] In the same vein as the environmental determinism of Montesquieu and Rousseau,[66] *terroir* brings the abstract notion of character to the world of agriculture. One might expect to find the metaphor of *terroir* in Weil's texts; after all, she did work in the vineyards in Saint-Marcel-d'Ardèche, which might have made her more likely to think in terms of *terroir* and the relationship between *nourriture terrestre* and *nourriture spirituelle*. However, it is in Heidegger's work that the idea recurs in various guises. In his corpus, land shapes the temperaments of the people who inhabit it, and every poem or thought that an individual produces ideally contains in its essence the residue of a specific

65. Parker, "Saint-Evremond," 129.

66. In the chapter "On the Laws in Their Relation to the Nature of the Climate" in *The Spirit of the Laws*, Montesquieu famously argued that climate has an impact on the character and passions of people in each region and that law-making should take account of this fact. Rousseau, in his *Essay on the Origin of Languages*, writes, "Whether one investigates the origin of the arts or examines the first morals, therefore, one sees that everything is related in its first principle to the means of providing for subsistence, and as for those among these means that gather men together, they are determined by the climate and by the nature of the soil. Thus, it is also by the same causes that the diversity of languages and the contrast in their characters must be explained." Rousseau, *Essay*, 309–10.

locale. *Terroir* turns land into an abstract idea. This puts Heidegger's rootedness at odds with that of Weil.

Heidegger, the *philosophe campagnard*, is perhaps best known for his lasting influence on poststructuralist thought, for his affiliation with the National Socialists, and for his inclination toward rural life, symbolized by his legendary Black Forest hut in Todtnauberg.[67] His mixed legacy was stirred again recently upon the publication of the *Schwarze Hefte* (*The Black Notebooks*), which include flagrantly anti-Semitic passages, casting a shadow over what was left of his reputation among his contemporary proponents. Much of his postwar thought responded to the crisis of late modernity and called for Europeans to reground themselves to avoid the nefarious effects of new technologies. Charles Bambach argues that Heidegger's careful construction of his own sylvan myth began in the radio address "Creative Landscape: Why Do We Remain in the Provinces?" ("Schöpferische Landschaft: Warum bleiben wir in der Provinz?," broadcast in 1933 and published in 1934) and continued through his well-known essay "On the Origin of the Work of Art" ("Der Ursprung des Kunstwerkes," [1935]), his 1949 essay "Feldweg," and his 1950 book *Holzwege*, up to his *Gelassenheit* address. One could add to this list "On the Essence of Ground" ("Vom Wesen des Grundes" [1929]), an early iteration of his notion of groundedness. Heidegger opts most often for the term *Bodenständigkeit*, which translates alternately as rootedness, groundedness, or autochthony, instead of *Verwurzelung*, the German equivalent of Simone Weil's term *enracinement*.[68] The *topos* of groundedness can be found already in some of his earliest writings and is pervasive in his texts between 1933 and

67. See Sharr, *Heidegger's Hut*.

68. Heidegger does use the term *Verwurzelung* occasionally in his writings. For example, he questions the possibility to move from *Entwurzelung* (uprooting/uprootedness) to *Verwurzelung* (rooting/rootedness) in this passage:

> Mastery over the masses who have become free (i.e., rootless and self-seeking) has to be established and maintained with the fetters of "organization." Can what is organized in this way grow back into its originary soil—not only blocking what belongs to the masses but *transforming* it? Does this possibility still have any chance at all in the face of the growing "artificiality of life," which renders easier and itself organizes that "freedom" of the masses, the arbitrary accessibility of all for all? No one should underestimate the importance of standing up to and resisting the unswerving uprooting. That is the first thing that must happen. But does that—and above all the means necessary to achieve it—guarantee the transformation of the uprootedness into a rootedness?
>
> Heidegger, *Contributions*, 43

1945, during his affiliation with National Socialism. Jeff Malpas has dedicated an entire book to Heidegger's treatment of place, arguing that this omnipresent theme in his work often takes form in a wide range of topological terms, such as *Ort* (place), *Ortschaft* (locality), *Stätte* (site, location, place), *Gegend* (region, area, district), *Dasein*, and *Lichtung* (clearing). He argues that Heidegger should be considered a founder of "place-oriented thinking" and divides his treatment of place into three categories,[69] "as the proper *focus* of thinking," "as that which is the proper *horizon* of thinking," and "as that which is the proper *origin* of thinking" (13). This last category is of most interest when considering the problem of Heidegger's *terroirisme*,[70] given that in his view, thought and all forms of cultural production emanate from the land and are marked by it. Robert Mugerauer has noted the pervasive theme of homecoming present in Heidegger's late texts, to which he attributes a very specific vocabulary, including *Heim* and *Heimat*, homelessness, dwelling, site, place, *Geheimnis* (mystery or secret), *Bodenständigkeit* (groundedness), and *Bodenlosigkeit* (groundlessness).[71] Recalling Bachelard's taxonomy of images of repose, the home and the root are first and foremost figures of return. Of Heidegger's concept of *Dasein*, George Steiner writes that Being is necessarily situated in the real world, claiming that "to be human is to be immersed, implanted, rooted in the earth, in the quotidian matter-of-factness of the world ('human' has in it *humus*, the Latin for 'earth')."[72] But Heidegger's is not an *irgendwo* or "just anywhere" embeddedness, which leads us to the first distinction between him and Weil. Weil's is a rootedness of generality; Heidegger's is a rootedness of specificity.

He delivered a memorial address called "Gelassenheit" (Releasement) on October 30, 1955, in his home village of Meßkirch on the occasion of the 175th birthday of the local composer Conradin Kreutzer. Charles Bambach argues that in this address, he was attempting to distance himself from the National

---

Despite his occasional use of *Verwurzelung*, the term *Bodenständigkeit* is much more illustrative of his nuanced connection between the human and the earth.

69. Malpas, Heidegger's *Topology*, 1.

70. Malpas explains what he means by "place as proper origin of thinking" in Heidegger's work: "[A] key point around which his thinking constantly turns is the idea that thinking arises, and can only arise, out of our original encounter with the world—an encounter that is always singular and situated, in which we encounter ourselves as well as the world, and in which what first appears is not something abstract or fragmented, but rather the things themselves, as things, in their concrete unity." Ibid., 14.

71. Mugerauer, *Heidegger*, 4.

72. Steiner, *Martin Heidegger*, 83.

Socialist tropes of his earlier writing.[73] While the audience may have expected the illustrious philosopher to discuss Kreutzer or music in general, they received instead an hour-long address about possible responses to technology in the atomic age. He connects this theme to Kreutzer by first broaching the subject of the composer's geographic origins, shared by Heidegger himself, located there in the south of the Baden-Württemberg state in Germany.[74] He poses a rhetorical question to his audience, to which he already has an answer: "Does not the flourishing of any genuine work depend upon its roots in a native soil?"[75] Heidegger's hypothetical question is a clear reference to Friedrich Schleiermacher's claims about the rootedness of a work of art in its place of origin. Schleiermacher wrote in the early nineteenth century: "Hence a work of art . . . is really rooted in its own soil, its own environment. It loses its meaning when it is wrenched from this environment and enters into general circulation; it is like something that has been saved from the fire but still bears the burn marks upon it."[76] To bolster his affirmative response to this question, Heidegger quotes Johann Peter Hebel, a writer who spent much of his life in Baden as well: "We are plants which—whether we like to admit it to ourselves or not—must with our roots rise out of the earth in order to bloom in the ether and to bear fruit."[77] Hebel clearly echoes Plato's *Timaeus*, although the human here is rooted in the earth but rises to the ether, or the heavens, and transcendence means the bearing of fruit, ostensibly the fruit of creativity and poetic production. The address then goes on to lament the fact that after the war, many Germans fled, and those who remained were figuratively displaced through television and radio. He ascribes this detachment from the *Heimat* to a general state of rootlessness throughout the country and as a general condition of modernity. Heidegger's reactionary technophobia is evident throughout the text; however, what is unique here is his attempt to formulate a

73. "I will look at Heidegger's discourse about the homeland, roots, landscape, and the poetry of Johann Peter Hebel as constituting a rustic form of politics whose meaning has been missed. I hope to show that the very topos of the homeland became a covert kind of political engagement for Heidegger that repositioned the old National Socialist philosophy of roots and soil in the new political context of a cold war Germany 'caught in a pincers' between America and the Soviet Union." Bambach, "Heidegger, Technology, and the Homeland," 268.

74. Ernst Bloch noted the tendency in this region toward the kind of paganism celebrated by National Socialists: "The whole of Alemmani-Bavaria is obsessed with the earth, and its writers and ideologists all the more so the more Christian they seem." Bloch, *Heritage*, 50.

75. Heidegger, "Memorial Address," 47.

76. Schleiermacher, *Ästhetik*, 86.

77. Heidegger, "Memorial Address," 47.

healthy response to technology, one that will provide the conditions for a new groundedness without rejecting technology altogether. He begins an interpretation of Hebel's words :

> The poet means to say: For a truly joyous and salutary human work to flourish, man must be able to mount from the depth of his home ground up into the ether. Ether here means the free air of the high heavens, the open realm of the free spirit.
>
> We grow more thoughtful and ask: does this claim of Johann Peter Hebel hold today? Does man still dwell calmly between heaven and earth? Does a meditative spirit still reign over the land? Is there still a life-giving homeland in whose ground man may stand rooted, that is, be autochthonic? (47–48)

Again, in Heidegger's reading of Hebel, we find that the path to the heavens is contingent upon rootedness, this time in regionally specific soil. If one loses this capacity to infuse one's work of art or even one's life with evidence of its regional specificity, one's origins are severed and transcendence through art is no longer possible. What he describes here is a notion similar to the French *terroir*, the term used in winemaking and other kinds of agriculture to identify the specific characteristics of flavor and aroma imparted to the product by local conditions such as soil, climate, and water supply, which differ greatly from one field to the next in some regions.[78] While acidity, color, sugar content, and other attributes of a particular bottle of wine are certainly altered by the land and its weather, the conceptual resonance of *terroir* echoes far beyond chemistry. The vintner knows it is a transcendental celebration of regional specificity. The word isn't always positively connoted; *sentir le terroir* means to embody either the qualities or the flaws of regional flavor.[79] The negative meaning expresses a tension between city and country, between cosmopolitanism and provincialism, in which people from the country see themselves as more authentic than those from the towns, and the urbanized see the rurals as ill-bred and culturally backward or belated. The idea of *terroir* has gained cultural capital in recent years, particularly in France, as producers of locally

78. The *Grand dictionnaire universel du XIX siècle* defines *terroir* simply as "Terrain considéré au point de vue de la production agricole." Larousse, *Grand dictionnaire,* 1660.

79. The *Grand dictionnaire* defines the expression *sentir le terroir* this way: "Avoir les qualités et surtout les défauts du pays où l'on est né ou que l'on a habité: *C'est un Gascon qui sent bien le terroir. L'auteur est Auvergnat et son livre sent le terroir.*" Ibid. It is worth noting that in Italian, the word *terrone* is a derogatory term used by northerners to describe provincial southerners.

grown agricultural products have sought better ways to compete against large-scale manufacturers of heavily processed foods. To describe the disconnection between people and their food, Olivier Assouly has proposed the term *déracinement alimentaire*, the uprootedness from gastronomic traditions of their region, people's lack of knowledge about the food they eat, and the inability to cook.[80] Relying on a nostalgia for pastoral life, the local producers have attempted to claim that their products are more authentic than factory-processed food, which is tainted with preservatives, artificial colors, and flavor enhancers. Their products, they argue, are more *real*. What these local producers claim about food, Heidegger claimed about art. Only land can provide the stamp of authenticity on vegetables and verses.

As Matt Kramer has argued, "[*Terroir*] sanctions what cannot be measured, yet still located and savored. *Terroir* prospects for differences. In this it is at odds with science, which demands proof by replication rather than in a shining uniqueness."[81] Heidegger is a difference-seeker, but in the realm of "what cannot be measured." The ineffable qualities that distinguish the home-grown from the counterfeit are very effective tools for making an unlosable argument. Similar to Heidegger's evocation of imagined etymologies in order to reinforce a particular claim, the logic of *terroir* asks the recipient of his argument to trust him and his authenticity radar. He relies on a certain *je ne sais quoi* or a *soupçon* of something in a poem, something in the air, something in a glass of wine that others must simply believe that he perceives though they cannot. In contrast to Weil, he celebrates the abstract qualities of *Heimat*, perceptible only by a native son. The abstract is what helps him give credence to his claims: untraceable etymologies fashioned into vague evidence, local textures, and newly coined metaphysical concepts all mobilize abstraction for the purpose of reassurance. Less than the abstracting tendency of modernity, he feared uniformization; he needed to believe in the singularity of his experience. However, distinction translates easily into separate but unequal; despite Heidegger's attraction to abstraction, Claudio Magris points out in his *Danubio* that the philosopher lacked the imagination to understand that the experiences of someone he'd never met were as authentic as his own.[82] In his engagement with National

80. Assouly writes, "Appliquée à la cuisine, la métaphysique n'a pas à jongler avec les apparences, mais à révéler l'essence des choses, préférant la nature à l'artifice, l'original à la copie, l'identité à la différence, la réalité à l'illusion, l'authentique au simulacre." Assouly, *Les nourritures nostalgiques*, 43.

81. Kramer, "Notion of *Terroir*," 225.

82. Magris, *Danube*, 45–46.

Socialism, Heidegger proved himself incapable of empathic projection. His *terroir* was the only legitimate *terroir*. Karl Jaspers, who denounced Heidegger's affiliation with the Nazis, conceived of German guilt in terms of a rootedness in the "German spirit and soul,"[83] into which a strong sentiment of coresponsibility could seep. Through the German cultural *terroir*, the nation was reaping the seeds it had sown.

More abstractly, *terroir* is a concept whose main purpose is to recontextualize. Not knowing where one's wine comes from, not knowing the environmental conditions that made it possible, is to take the product but leave its context behind, constituting a kind of dumb consumption. *Terroir* is a recuperative notion that tries to tell a resituating story, the origin story of a wine or—in Hebel's and Heidegger's case—of a person. For example, the *Appellation d'origine contrôlée*, the French certification process with strict rules for identifying the origin and regulating the naming of agricultural products, finds its human equivalent in the issuing of birth certificates, going one step further in its insistence on a correspondence between name and place. The creation of a bureaucratic framework through which to regulate something as slippery as *terroir* is no easy task. When the concept is taken up by philosophy, it becomes even more dubious. Because of its approximative nature and quasi-spiritual implications, *terroir* proves very useful in Heidegger's volatile argumentation. While Weil's and Heidegger's searches for context overlap in many ways, Weil does not rely on *terroir* to make her case. Her approach to recontextualization is more pragmatic and—strangely—less mystical than his. In this instance, Heidegger proves to be much more of a mystic than Weil.

Heidegger's speech goes on to lament the loss of *Bodenständigkeit*, or groundedness, caused by the spirit of the age. While the term can be translated as "rootedness" or "autochthony," the translation "groundedness" is preferable because the root *Boden* here means "ground" or "soil." The term *Ständigkeit* alone may mean constancy, steadfastness, stability, or permanence; thus, attaching *Boden-* as a prefix implies terrestrial fixedness or sureness of earthly foundation. The adjective *bodenständig* is a part of everyday language in German, both in Heidegger's time and now. It means down-to-earth or humble and is used to describe people. In a signature move, Heidegger attempts to

83. "Thus the German—that is, the German-speaking individual—feels concerned by everything growing from German roots. It is not the liability of a national but the concern of one who shares the life of the German spirit and soul—who is of one tongue, one stock, one fate with all the others—which here comes to cause, not as tangible guilt, but somehow analogous to coresponsibility." Jaspers, *Question*, 73.

estrange the familiar by forcing his audience to approach the mundane word with fresh eyes, to notice the ground of which it is composed, and to literalize and transform it into a philosophical concept. He gives new attention to *Bodenständigkeit*, thus making it visible again and showing the extent to which the contemporary world lacks it. For Heidegger, the groundlessness of life in his time was a product of the very age itself, not of a particular behavior or laxity. More specifically, he refers to the incommensurability of the human with technology. He proposes a solution, what he calls *Gelassenheit*, a term borrowed from the medieval theologian and mystic Meister Eckhart. The word had the earlier metaphysical connotation of giving oneself over to God, but Heidegger secularizes it.[84] For him, *Gelassenheit* involves a relaxing of the self when faced with any unfamiliar and potentially threatening phenomenon, a giving-oneself-over to the object at hand. This counterintuitive relaxation before the unknown offers, in his view, a way to rectify the harmful pattern he'd already identified in the preatomic age and to prevent its transmission to the nuclear future. He elaborates on the relationship between *Gelassenheit* and *Bodenständigkeit*: "Releasement toward things and openness to the mystery give us a vision of a new autochthony which someday even might be fit to recapture the old and now rapidly disappearing autochthony in a changed form" (55). He ends with this hypothetical scenario, thus returning, if only obliquely, to the occasion for which the memorial address was presented: "If releasement toward things and openness to the mystery awaken within us, then we should arrive at a path that will lead to a new ground and foundation. In that ground the creativity which produces lasting works could strike new roots" (56–57). So *Gelassenheit* and openness toward the mystery become for him a panacea not only for the anxieties about the uncertain atomic future but also for the perceived loss of regionally rooted creativity.

One can discern in this strange memorial address Heidegger's attempt to redefine the theme that had dominated so much of his earlier writing, that of local specificity, which, given the context in which that writing was produced, is inextricably bound to toxic nationalism. His celebration of the local involves an emphasis on the appreciation of all of the bioregional features of one's homeland, which have an impact on culture. The earthiness of Heidegger's project, from his fascination with *Holzwege* (paths made for seeking wood) and the *Feldweg* (field path) to the mystification of his Black Forest hut, shares with the French notion of *terroir* an attachedness to one's land of origin, a

84. The translators suggest that *Gelassenheit* also means "composure," "calmness," and "unconcern." Heidegger, "Memorial Address," 54n4.

rootedness not only to the land's beauty and bounty but to its capacity to distinguish the authentic, local, and good from the inauthentic, foreign, and bad. One's roots mark not only the work of art, but any type of thought, as he asserts in this declaration: "I am convinced that there is no essential work of the spirit that does not have its root in originary autochthony."[85] A national myth can easily be forged by binding *Volk* and *Boden*; any human-as-citizen can claim an attachment to the ground (national territory) on which the person dwelled since childhood.[86]

Furthermore, the sacralization of one's own dwelling space creates the conditions for a self-imposed custodianship of the plants, the water, the soil, and the creatures that dwell there. This is why so many scholars have framed Heidegger as an ecological philosopher. It is essential to recall that the twentieth-century ecological movement in Germany was originally bound to far-right politics. This right-wing German environmental preoccupation is clearly a pagan ecology. Unlike the United States and other European nations, where ecological interests were originally left-leaning, in Germany it was largely the National Socialists who advocated the stewardship of the land, which was seen as the embodiment in nature and the progenitor of the German *Geist*. Indeed, the paganistic obsession with the earth's earthiness in whatever form seems to be a signature of Germanic culture.

Like Heidegger, Jung expressed his fear of human alienation from nature and tradition in the form of a critique against America, whose daily way of life he saw as, "in a subtle way, so *déraciné*, uprooted."[87] Europe's fear of American modernity, so distant geographically and philosophically from the slow-simmering developments of the old country, takes form most often as a lament against the self-orphaning tendency of the United States. The dismissal or destruction of nature there and the alienating aspects of work and home life,

85. Heidegger, *Reden*, 551. "Ich liebe dies alles, weil ich überzeugt bin, daß es kein wesentliches Werk des Geistes gibt, das nicht seine Wurzel in einer ursprünglichen Bodenständigkeit hat." Translation by Charles Bambach, *Heidegger's Roots*, vi.

86. Adam Sharr raises similar questions in his *Heidegger's Hut*: "Charges of invidiousness may always attend any romance of a lost provincialism supposedly free from cosmopolitan delusions. Where some sort of authentic rootedness is claimed, must there not be outsiders inevitably doomed to inauthenticity? . . . Where the transcendence of 'nature' is evoked, might it not allow an unhealthy detachment from human responsibility? Moreover, might not biological determinism and the rhetoric of blood and soil follow close behind?" Sharr, *Heidegger's Hut*, 108–9.

87. Jung, *Earth*, 143.

consumerism, and technological innovation were signs that these children of Europe wanted to deliberately cut themselves off from Mother Nature and the Fatherland. Jung saw in Europe's adoption of "the American tempo" a sign that his own continent could not resist the tendency to self-eradicate at the same pace as the United States; he conceived it as "a final race between aging Europe and young America" (123). Nature gets caught in the crossfire of the worldwide competition for technological advancement, but, as Jung asserts, severing the tie with nature is a first step toward human exile. He claims that "natural life is the nourishing soil of the soul" (67),[88] an image that transforms the soul into a plant.

Early in the twentieth century, the Austrian philosopher Rudolf Steiner developed a school of thought known as anthroposophy, which involved the spiritual rerooting of the human in nature, work, education, family life, and all aspects of daily existence. Steiner was the founder of Waldorf Education, which has an international presence today, and his entire pedagogy relies on the root metaphor.[89] From the plant, the child learns one of the central tenets of anthroposophy: that human life should take form as cosmic embeddedness.[90] In endless comparisons between plants and humans, Steiner comes to the conclusion that humans are truly botanical beings.[91] Unsurprisingly, he also

88. He uses the metaphor again and again in his writings, for example here: "For it is the body, the feeling, the instincts, which connect us with the soil. If you give up the past you naturally detach from the past; you lose your roots in the soil, your connection with the totem ancestors that dwell in your soil. You turn outward and drift away, and try to conquer other lands because you are exiled from your own soil." Ibid., 73.

89. "The flower and fruit of a plant exist in the root, and if the root receives proper care, both flower and fruit will develop properly with sunlight and warmth. Likewise, soul and spirit live in human bodily nature, the body created by God. If we take hold of the body's roots—knowing that divinity lives there—and develop them correctly, surrendering to the life that is freely developing, then the soul and spirit in those roots develop, as do the inner forces of a plant, which pour from the root and develop in the warmth of sunlight." Steiner, *Modern Art*, 26.

90. "The roots of the plant are intimately connected to the earth. As for plants, the only thought we should awaken in children is that earth and root belong together, and their only thought about blossoms should be that they are drawn from the plant by sunlight. Children are thus led out into the cosmos in a living way." Ibid., 127. "[W]e explain how sand and stone began with what was once destined to become plant, how the earth is like one huge plant or giant tree, out of which the various plants grow like branches" (129).

91. "When we look at a plant, something is already contained there in the seed that takes root and, after a long time, will appear as blossom and fruit. Similarly, in children before the change of

looked toward the Greeks as an influence,[92] and he took interest in what he calls the "root races" of humanity in his book *Cosmic Memory: Prehistory of Earth and Man*. Unlike Heidegger's nationalizing or regionalizing ontology, Steiner's reflections on origin and terrestrial embeddedness resemble Simone Weil's, focusing more on the holistic spiritual well-being of the individual than on any geographically specific autochthony. As shown in chapter 2, anthroposophists like the poet Christian Morgenstern took up Steiner's metaphor of rootedness and essentially canonized it as one of the central tropes of the movement. The National Socialists denounced Steiner and his movement despite its obvious racializing undertones and its paganistic sacralization of nature. Steiner seemed to favor a postracial, postidentarian civilization in which race and nation would be absorbed into a universal connectedness. A clear genealogy can be traced through early Germanic paganism, its residue in the Middle Ages in legends like the mandrake story, the botanical obsessions of the German romantics, right-wing environmentalism, anthroposophy, and Heideggerian land worship.

In his text "Heidegger, Gagarin and Us," Emmanuel Levinas describes Heidegger's attachment to the earth as a form of infantile nostalgia and paganism.[93] The dangerous pagan myth of earthly embeddedness appeals to the childish part of the self that resides in the mystery, experiencing it with total fascination. He believes that Judaism understood something that other religious cultures have not, namely that destruction of sacred groves symbolizes an elimination of the mystery that motivates humans to be cruel to one another.[94] He addresses Heidegger's technophobia, noting that technology is less dangerous in the end than the proponents of place because "place" is most often a divisive concept

---

teeth, when the bodily nature is susceptible to the soul's influences, there are seeds of happiness and unhappiness, health and sickness, which will affect all of life until death." Steiner, *Roots*, 10.

92. "The Greeks' expectation of the body was exactly what we expect of a plant: that, if the root has been treated properly, it will blossom on its own under the influence of sunlight and warmth." Steiner, *Modern Art*, 20.

93. "Enrootedness. We should like to take up this term; but the plant is not enough of a plant to define an intimacy with the world. A little humanity distances us from nature, a great deal of humanity brings us back. Man inhabits the earth more radically than the plant, which merely takes nourishment from it. The fable spoken by the first language of the world presupposes links that are more subtle, numerous and profound." Levinas, *Difficult Freedom*, 232.

94. "This, then, is the eternal seductiveness of paganism, beyond the infantilism of idolatry, which long ago was surpassed. *The Sacred filtering into the world*—Judaism is perhaps no more than the negation of all that. To destroy the sacred groves—we understand now the purity of this apparent vandalism. The mystery of things is the source of all cruelty towards men." Ibid.

that separates people into natives and foreigners. In his view, technology affords the opportunity to do away with the stubborn mysticism that represses the human. Levinas then addresses the stereotype that the city, the perceived locus of the modern Jew, is where humans lose their connection with nature. He traces a lineage between Socrates and the Jews, claiming that Socrates championed the city as the space of the social and understood that this human congregation was essential for thought. Levinas goes on to argue that Christianity and Judaism differ the most in their definition of the human's relationship to the earth. The former nurtures "family, tribe and nation" while the latter "has freed Nature from its spell" (233–4). Maurice Blanchot shares this conviction, arguing that fixedness is a primary characteristic of paganism and that the Jew, who resists this immobility, is more prepared for the contingencies of life, which might force one to move at any moment. For him, "to be pagan is to be fixed, to plant oneself in the earth, as it were, to establish oneself through a pact with the permanence that authorizes sojourn and is certified by certainty in the land."[95] Judaism wants to break this pact and to do away with the geographic certainties. Levinas pursues this argument in his article "Philosophy and the Idea of Infinity" as well, claiming that the "earth-maternity determines the whole Western civilization of property, exploitation, political tyranny, and war."[96] The fact that pagans, the National Socialists, and Heidegger take embeddedness as the default and "natural" condition of the human is the most deleterious aspect of this worldview. Levinas's claim becomes problematic in light of Zionist efforts to reroot Jews in a homeland; the tension in contemporary Judaism between a conscious celebration of nomadism as a chosen form of dwelling in the world and the celebration of a recovered nation shows that the view of Jews as "naturally" rootless people is extremely reductive.[97] Levinas's denunciation of Heidegger's thought goes far beyond him as an individual; rather, he is a mere representative of a widespread worldview determined by the centrality of place in one's personal mythology. As I show in chapter 6, what Levinas describes here constitutes what Stephen Pepper has

95. Blanchot, *Infinite Conversation*, 125.

96. Levinas, "Philosophy and Infinity," 53.

97. I'd like to thank Maurie Samuels for helping me to understand the nuances in this debate. Examples of the celebration of Jewish nomadism include Vilém Flusser's "Taking Up Residence in Homelessness" and Sarah Hammerschlag's *The Figural Jew*. Examples of the celebration of Jewish rootedness include most obviously all of Theodor Herzl's texts on the establishment of a Jewish state and recent books such as Shmuel Trigano's *Politique du peuple juif: Les Juifs, Israël et le monde* (Paris: Bourin, 2013); and Caryn Aviv and David Schneer's *The New Jews: The End of the Jewish Diaspora* (New York: New York University Press, 2005).

called a root metaphor or ground metaphor, that is, a basic metaphor, usually assumed early in life, that determines all other aspects of one's worldview. For Heidegger, Levinas argues, the entirety of his philosophy departs from a basic false belief that home and land exert a quasi-mystical influence in the formation of person as a culture-producing subject.

*

One finds many parallels between Europe's situation today and that of Weil's and Heidegger's Europe. New technologies continue to provoke anxiety. Nuclear energy, drones, genetically engineered food, and surveillance technologies like facial recognition and metadata-collection software keep people on edge. There is particular concern over the progressive zombification of the human through rampant use of smartphones and other digital technologies that turn social interaction into a remote event. Europe has witnessed a resurgence of ultranationalism in places like Greece (Golden Dawn), France (Marine Le Pen and the new generation of the Front National), Germany (Pegida and the Nationalsozialistischer Untergrund), and Italy (Forza Nuova, Fiamma Tricolore). The increased intensity of Islamophobia has manifested itself in campaigns against the building of mosques, the wearing of the headscarf in public schools, and the implementation of Sharia Law on European soil. Furthermore, new political parties have emerged whose primary objective is to dismantle the European Union. The demands of capitalism and consumption culture have detrimentally impacted not only the environment but the dignity of the working class, whose jobs are constantly at risk, either for export owing to the lack of an international minimum wage or because mechanized means of production make human workers unnecessary. The university has become more a career-development hub than an institution for the development of critical minds and a well-rounded set of skills for living; the bureaucratization and commercialization of the university, more in the United States than in Europe, have created a student body more interested in crafting a flawless C.V. than in fully assimilating ideas and nurturing a comprehensive understanding of the mechanisms of politics, history, aesthetics, and other fields. There is just enough comfort and distraction to keep the masses pacified and reliant on a steady stream of new needs, and thus to forestall an insurgence. How would Weil and Heidegger respond upon discovering that the problems they anticipated have impacted future generations to such a vast degree?

Scrutinizing Weil's *The Need for Roots* reveals that her main interest seems to lie in the daily experience of working-class and rural people rather than the promotion of one nationality, race, ethnicity, or creed over another. Greatest in numbers, the working class is the necessary foundation upon which a stable society must rely. Weil was cured of her undying patriotism for France at a young age, after the punitive effects of the Treaty of Versailles became apparent. She discovered that every nation, every people, every individual is fallible. While she does promote cultural continuity through the nurturing of traditions, she does not believe that the ancestors were infallible, either. As David McLellan writes, "[Weil] did not share Barrès's worship of the earth and the dead and her conceptions of hierarchy and order were far from those of Maurras and the *Action Française*."[98] She was no cultural purist; her thought is the result of an active combinatoric practice, fusing a wide variety of influences and testing the possibilities and limits of each. She critically interrogated the influences on her thought. For example, she challenged many aspects of Leftist activism and Christianity even while espousing most Leftist or Christian tenets. While Weil undoubtedly approaches the question of rootedness from a French Catholic perspective, her vision involves a more widely applicable set of principles that anticipate many of the most pressing problems of *post*modernity. From vocational estrangement to total alienation from the natural environment, from the abysmal cleft between education and real life to abridged cultural memory, Weil specifically addressed, already in the 1930s and '40s, these problems that afflict us today. Despite her deployment of a mystically fervent vocabulary, the content of the better part of her recommendations is secular and devoid of racial or national implications. Several aspects of Weil's understanding of rootedness are surprisingly relevant today. Though she explicitly addresses Europe, the same solutions for providing humans with a sense of belonging in the world could be equally relevant in China, Angola, or Ecuador. In the passages that get away from the specificity of Europe and address the estranging aspects of international modernity, or rather anational modernity, her prescriptions insinuate a borderless community of workers, farmers, teachers, and people of all vocations and ages; in short, she addresses citizens of the world. She uses France and Europe as case studies to illustrate a more general problem of human detachment provoked by the conditions of modernity. I believe Weil correctly identified the real cause of many of Europe's woes in the twentieth and twenty-first centuries. Abstraction, not simply technology, is what could lead to a catastrophe of global magnitude.

In Heidegger's case, the fear of estrangement from one's surroundings owing to new technologies certainly resonates today, particularly in relation to the

98. McLellan, *Utopian Pessimist*, 257–58.

Internet, digital telephony, and media infatuation. His description of a popular addiction to alienating devices could not be more applicable than to the now. The 1980s through the present have been marked by a fear of digital technology; the language of pathology (the virus, the worm), the language of warfare (spyware, Trojan horses), and other ominous terminologies (ransomware, rogue security software, the bug) to describe malevolent programming practices are indicative of this fear. The *Gelassenheit* address promotes a moderate response to the new, encouraging people to subdue the fear reflex rather than staunchly refuse all technological innovations. As an individual, one cannot prevent the arrival of new technologies; one can only temper one's response to them. Heidegger's recommendations are relatively abstract and expressed through the quasi-mystical vocabulary of release in the style of Meister Eckhart. Such a vocabulary implies that the cold objectivity of Enlightenment reason is insufficient in responding to the new technological advancements for which it was in part responsible. The Enlightenment hoped to shine a glaring light on every dark corner of civilization. It hoped to expose the mystery. Heidegger seems to propose in his *Gelassenheit* address a reenchantment of life and a nurturing and acceptance of the mystery without the need to lay it bare. By unstiffening one's response to the unknown, one begins to move in the same inevitable direction as the flow of the new rather than resisting this current to no avail. This is not to say that Heidegger was suddenly a Zen master who had emptied himself of Self; he remained warily critical of the human's relation to technology. Still, he recognized the futility of categorical refusal. Fear of the unknown takes other forms as well. For example, one could place the fear of a new technological device in the same category as the fear one feels toward a neighboring nation about which one knows very little. Heidegger recognized his own anxieties toward the new and foreign in general and their unknown consequences, as what is new or foreign is inherently *unursprünglich*, far from the origin, far from home. Opposing the impending danger of nuclear proliferation, Heidegger's late iteration of *Bodenständigkeit* takes as its implicit adversaries the United States and the USSR, allowing him to express fear of two overlapping unknowns simultaneously: the cultural influence of these world powers and the destructive technologies they wield. Fear of the alien neighbor brings Heidegger back to where he feels most at home: celebrating the knownness and knowability of his own region. Through the logic of *terroir*, Heidegger resists uniformization and makes the case for regional specificity. The weather, topography, and the qualities of the soil and water produce *difference* in plants and people that originate in a particular spot. To cut ties with these difference-imparting phenomena is to accept a future of sameness.

If Weil and Heidegger had been asked to take sides in the *Querelle du peuplier*, I feel confident that Heidegger would have sided with Barrès in his regionalist appeal, which borrows a *terroir*-like logic to prove that land leaves its mark on people. Celebrating the specificity and perfection of one's place of origin usually demotes someone else's homeland. The *terroir* logic rarely involves rankless specificity; its focus is qualitative difference. If you can imagine a conversation between Heidegger and Barrès about which is better, a small corner of the Black Forest or a small corner of Lorraine, the debate would certainly be contentious. This kind of exclusive particularism and the search for difference can create an atomized and volatile politics. As for Simone Weil, while she does share Barrès's fear of intellectualism and abstract reason, I don't believe she would have been particularly inspired by either his position or Gide's. Her singular take on the question of rootedness does not line up neatly with the typical discourses of nation, race, nature, or religion. Her diagnosis of Europe's problems is holistic, informed by a strong understanding of history, an amalgamated spirituality, and an obstinate belief that abstraction is at odds with human prosperity. If we were to embrace Heidegger's call for a new *Bodenständigkeit*, we would be following the same logic of *terroir*, which has a difficult time acknowledging the mystical element behind it. The contemporary fetishization of locally produced food and wine represents he attempt to resist the corporate invasion of agriculture while still participating in the consumerist compulsion. To turn land into a mystical rootbed that infuses human cultural production with its magic seems in many ways a more exploitive idea than for a multinational food corporation to harm the environment in order to make a profit. When the homeland itself becomes a kind of exclusive brand with supernatural powers, of which everyone wants a piece, the political and social consequences can be devastating. Heidegger's abstraction of the land into mystically infused *terroir*, which cannot be quantified, participates in the very mechanism that Weil feared. At least she is open about the mystical elements in her thinking. Her plea for a new rootedness in an unspecified location—whose historical traditions should be learned, whose fauna and flora should be studied, and whose contours should be contemplated—asks everyone to embrace the place where he happens to be. Polynesian children should not be forced to recite "our ancestors the Gauls";[99] they should know about local traditions, spiritual practices, agriculture, and the necessity of their vocations. In her idealized conception, one culture should not be celebrated to the detriment of another. However, this utopianism is complicated by her

99. Weil, *Need for Roots*, 46.

fear of abstraction, which led her to attribute the greatest cause of European rootlessness to capitalism, intellectualism, and Judaism. This sweeping verdict undermines the more powerful parts of her argument, which undoubtedly could be of service in the contemporary world, in which abstraction continues to triumph.

In the next chapter, Jean-Paul Sartre—along with other phenomenologists—takes up where Heidegger left off, trying to reground the human in this fact called Being. Sartre's special relationship with the tree as a real thing and as an idea informs his treatment of the root as a revolting and insistent reminder of existence, which stuns the protagonist at the foot of a chestnut tree in his novel *Nausea*. The root as the site of Roquentin's existential crisis was not an arbitrary choice; Sartre had in mind Descartes's Tree of Philosophy, whose roots are metaphysics. The upheaval of Western metaphysics is, for Sartre, the clear objective of his ontologically radical project. His philosophical investments were not in the land, in culturally or genetically transmitted heritage, or in the problem of primordial origin; while Weil and Heidegger could be described as retrophilic thinkers, with their admiring eyes fixed on the past, Sartre preferred the immediate. His rootedness, similar to the Buddhist's, was in the here and now, in the fact of Being.

CHAPTER 5

Sartre, Phenomenology, and the Root

Man is but a reed, the most feeble thing in nature; but he is a thinking reed.

BLAISE PASCAL, *The Thoughts, Letters, and Opuscules of Blaise Pascal*

"Philosophy can take root only in radical reflexion upon the meaning and possibility of its own scheme": This was Edmund Husserl's formulation to describe the aims of phenomenology.[1] Alternating between root, ground, and foundation, European phenomenologists of the twentieth century tried out various metaphors to describe their central claim, namely that the consciousness of phenomena always happens through a perceiving subject. Thought and perception have a "home base," so to speak, but a self-veiling one; the elision of this fact, perhaps at one time the most common elision in philosophy, prevents a full account of the way a thought is formulated. In an effort to underscore this fundamental fact, Husserl, Heidegger, Sartre, Merleau-Ponty, and others deploy the root metaphor specifically to describe how a better understanding of consciousness from a first-person perspective can be reached by turning to a seemingly reliable source of experience: the phenomena themselves as discerned by the subject. They hoped to reembody or reroot consciousness in a stable locus: the perceiving mind.

This chapter begins with a close reading of Sartre's *La nausée*, with particular focus on the famous root scene, which gives a radical form to the troubling impenetrability of existence. Sartre's choice of the metaphor was not arbitrary; the root and the tree hold a highly charged personal significance for him, as I will show in his letters, journals, and philosophical texts such as

1. Husserl, *Ideas*, 27. The various subcategories of phenomenology (including genetic, hermeneutical, realistic, generative historicist, existential, and transcendental and naturalistic constitutive) all articulate the necessity for some kind of philosophical grounding.

Being and Nothingness and *Truth and Existence*. Furthermore, Sartre's interest in the root and in "rooted phenomena" is consistent with the phenomenologists' ambitious project of (re)grounding philosophy as a discipline. I conclude with pertinent examples from the canon of phenomenological texts by thinkers like Husserl, Merleau-Ponty, and Heidegger, which reveal that their use of the root metaphor is not simply a rhetorically interesting way to say "origin." Their selection of a living, grounded form that connotes the essential quality of something is careful and meaningful. I show how rootedness and, alternatively, groundedness, are the organizing metaphors for phenomenological thought.

THE NAUSEA-INDUCING ROOT OF BEING

The most famous *mise-en-scène* of the root and its symbolic challenges to rational thought is found in Sartre's *Nausea*. The protagonist, Antoine Roquentin, settles in Bouville, which elicits through the French ear the image of a city of mud (*boue* + *ville*) or a city of the bottom or the end (*bout* + *ville*). This is the first of many abject cues; the novel displays wide-ranging forms of filth. As a historian conducting research on the life of the Marquis de Rollebon, Roquentin spends many of his waking hours in a veritable forest, the library, whose books, tables, and chairs are composed of wood. He lives alone and manifests the anxieties of the typical modern subject: he simultaneously fears alienation and the company of other humans; he feels his existence to be superfluous but continues to play his human role, each day going through the motions of societally programmed living; he finds culture and nature irreconcilable.

The botanical world imposes itself on the human throughout Sartre's narrative, even before the moment of illumination at the foot of the chestnut tree. In one passage, Roquentin studies his own face in the mirror and describes what he sees as "well below a monkey, on the fringe of the vegetable world."[2] He likens his own hair to blades of grass (137). He describes the cashier of the Café Mably as a "fleshy white plant" (78) seen externally through the café window. Fleshy plants under the rainy skies of Bouville take on the texture and form of human ears.[3] Violet—as a flower, a color, and a smell—is associated throughout the novel with loss, death, and the grotesque; it is the hue

2. Sartre, *Nausea*, 27, hereafter cited parenthetically in text and notes throughout this section of the text.

3. "They let plants grow between the gratings. Castrated, domesticated, so fat that they are harmless. They have enormous, whitish leaves which hang like ears. When you touch them it feels like cartilage" (209).

of unease and disease. Other flowers, such as the lily and the rose, also come to represent human death. As Bataille has argued, the flower is indissociable from sex, although the relationship between the two requires a displacement of the true locus of desire, the sexual organ, to a secondary site.[4] A young, deceased student of the Polytechnique in a portrait hanging in the Bouville museum is described as a cut rose, after a quotation from Anchises about Augustus's nephew Marcellus in Virgil's *Aeneid*: "*Tu Marcellus eris! Manibus date lilia plenis*" (translatable as "A new Marcellus shall arrive in thee" or "You will be Marcellus, give lilies with full hands").[5] The mushroom, a truly abject botanical form, inseparable from dirt and rot, also appears in Roquentin's pages. The narrator imagines his new life in Paris as a "mushroom existence" (231), and his former lover Anny describes the mildewed pages of the large tome of Michelet's *History* as having the same color and smell as a mushroom (196).[6] It is worth recalling that Freud used the form of the mushroom to describe the structure of a dream.[7] Roquentin also recalls Pascal's image of the human as a thinking reed.[8] He even describes human consciousness in botanical terms, writing in his journal, "Consciousness exists as a tree, as a blade of grass" (227).

The protagonist falls asleep into a monstrous botanical dreamscape while fondling the owner of *Rendez-vous des Cheminots*. This woman elicits disgust in him, as "she is too white and besides, she smells like a newborn child" (82). In *Being and Nothingness*, Sartre argues that "erotic descriptions insist on the smooth whiteness of a woman's body" because "what is smooth can be taken and felt but remains no less impenetrable, does not give way in the least beneath the appropriative caress—it is like water." Sartre explains, "The idea of

4. Bataille, *Visions of Excess*, 11.

5. "A cut rose, a dead polytechnician: what could be sadder?" (128).

6. The translator of the English edition omitted the line "elles sentaient aussi le champignon" ("[the pages] also smelled like mushrooms").

7. "The dream-thoughts which we encounter during the interpretation commonly have no termination, but run in all directions into the net-like entanglement of our intellectual world. It is from some denser part of this fabric that the dream-wish then arises, like the mushroom from its mycelium." Freud, *Interpretation of Dreams*, 368.

8. This English translation leaves out the Pascal reference "thinking reed" ("roseau pensant"): "What's the matter with your science? What have you done with your humanism? Where is your dignity?" (213). Pascal's original use of the term in context is this: "Man is but a reed, the weakest in nature, but he is a thinking reed. It is not necessary that the entire universe arm itself to crush him. A breath of air, a drop of water, suffices to kill him. But were the universe to crush him, man would still be more noble than that which kills him, because he knows that he dies; and the universe knows nothing of the advantage it has over him. Our whole dignity consists, then, in thought. . . . this is the principle of ethics." Pascal, *Thoughts*, 170.

'carnal possession' offers us the irritating but seductive figure of a body perpetually possessed and perpetually new, on which possession leaves no trace."[9] La Patronne, as she is called, has countless lovers, but the signs of their sexual encounters leave no trace. Roquentin is repulsed by this whiteness, the same color used to describe fleshy plants in the scenes mentioned above. As he falls asleep, the latent horror contained in her person comes alive in his subconscious:

> I played distractedly with her sex under the cover; then my arm went to sleep. I thought about de Rollebon: after all, why shouldn't I write a novel on his life? I let my arm run along the woman's thigh and suddenly saw a small garden with low, wide trees on which immense hairy leaves were hanging. Ants were running everywhere, centipedes and ringworm. There were even more horrible animals: their bodies were made from a slice of toast, the kind you put under roast pigeons; they walked sideways with legs like a crab. The larger leaves were black with beasts. Behind the cactus and the Barbary fig trees, the Velleda of the public park pointed a finger at her sex. "This park smells of vomit," I shouted. (82–83)

This passage highlights a *topos* in French literature, particularly since the period of the Decadents, of holding the botanical world in contempt as the vile force that encroaches upon the human world. In his book *Forests: The Shadow of Civilization*, Robert Harrison writes, "Nothing . . . disquiets a rationalist more than a forest."[10] Jean Pierrot takes a similar position, identifying what he calls "la revanche du végétal" (revenge of the plant world) in the French literary canon, particularly starting in the second half of the nineteenth century. Pierrot illustrates the ubiquity of the belief that plants work against human progress and the fear that they will eventually erase civilization. Offering a wide range of examples, starting with Emile Zola and ending with Roger Caillois, Pierrot argues that from the late nineteenth century onward, "the plant would be considered an essentially dangerous reality for man. It would appear to seek revenge for the long subjugation imposed on it by man, attempting to reverse in its favor, through the image of a rivalry between different species, the reign humans believed to have permanently established on Earth."[11] Behind the veil of threatening greenery, other hostile entities lurk. Ants, centipedes, and

9. Sartre, *Being and Nothingness*, 579.
10. Harrison, *Forests*, 146.
11. Pierrot, "La revanche," 252.

ringworms infest Roquentin's dream scene; these can bite, sting, or infect the human. The edible human is also the eating human, implied through the surreal image of the toast-shaped animals that move like crabs. In this moment, meaning breaks down completely, following the logic of the abject as defined by Kristeva.[12] The crab is a recurring figure throughout *Nausea* and is most memorable perhaps as Roquentin's own hand, which he perceives as detached from his body, having taken on a life of its own. This disconnectedness of the hand from the body is also prefigured here as Roquentin's arm falls asleep; the arm becomes sensorially detached from his body, a foreign object whose sensations are absent or at least distant. Roquentin dwells on descriptions of the hands of the Self-Taught Man (*l'Autodidacte*), a humanist who frequents the same library as Roquentin, which reach for the hands of young boys. In *Being and Nothingness*, in the section on bad faith (*mauvaise foi*), Sartre depicts the scene of a woman unwilling to decide whether she wants to sleep with a man who is courting her and describes the way she lets her hand transform into a thing detached from her body, which her suitor may fondle at will without any commitment from her person.[13] Roquentin's dream ends with the eruption from his throat of the mysterious phrase, "Ce jardin sent le vomi," a prefiguration of the scene in the Jardin public when the cause of his nausea becomes clear. The overtly sexual nature of this excerpt and its relation to the garden summon up images of the Fall of Man in the Garden of Eden.

Many abject scenes in the novel foreshadow Roquentin's illumination. One of the most striking of these is triggered when Roquentin buys a newspaper that announces that the raped body of a little girl named Lucienne has been found. Previously, as Roquentin passed through the Jardin public, he had noticed a man in a large cloak sitting on a bench, perhaps about to expose himself to a little girl around ten years of age. Roquentin unintentionally disturbs "their little drama" (109–10),[14] scaring the little girl away. The headline about

12. The abject has only one quality of the object—that of being opposed to I. If the object, however, through its opposition, settles me within the fragile texture of a desire for meaning, which, as a matter of fact, makes me ceaselessly and infinitely homologous to it, what is *abject*, on the contrary, the jettisoned object, is radically excluded and draws me toward the place—where meaning collapses. . . . On the edge of non-existence and hallucination, of a reality that, if I acknowledge it, annihilates me. There, abject and abjection are my safe-guards. The primers of my culture.

Kristeva, *Powers of Horror*, 1–2

13. Sartre, *Being and Nothingness*, 55–56.

14. This staging of a garden as the potential site for the loss of innocence has clear echoes in the Edenic garden. For a compelling account of the garden as a privileged site in the history

dead Lucienne may recall this earlier scene of sexual invasiveness. The name Lucienne is a variant of the name Lucie, meaning "light," used for another character in the book. Light, as Hans Blumenberg has shown, is a complex and widely used metaphor for the arrival of knowledge, and Sartre exploits this metaphor to the fullest.[15] Reflecting on Lucienne's case prompts Roquentin into a free association of abject images, which lead him closer to his own illumination. "Little Lucienne was raped. Strangled. Her body still exists, her flesh bleeding. *She* no longer exists. Her hands. She no longer exists" (136–37). The Cartesian *cogito* is undone in this passage: "I am. I am, I exist, I think, therefore I am; I am because I think, why do I think? I don't want to think any more, I am because I think that I don't want to be, I think that I . . . because . . . ugh! I flee" (137). The passage unravels into complete absurdity: "I do not think, therefore I am a moustache" (138). The dissolution of the Cartesian subject is at the heart of Sartre's narrative, and as I will show, it is not by chance that he chose the root as the site of this dissolution. Roquentin's string of associations continues to stage an interpenetration of subject and object, erasing the barrier between the one who violates and the one being violated. In the case of Lucienne's rape, one existence forces its way into another in the same way that Existence forces itself on Roquentin. Images of mud, fingers, rape, blood, bodily fluids, and "florescence of flesh" (138) lead to the culmination of the episode, an imagined rape from behind, *par derrière*. Anal sex means planting the sex in filth. In this instance, a penetration by force, the victim—victim because neither Lucienne nor Roquentin want the assault they receive—is surprised from behind, unable to see the coming attack. Anal violence is alluded to in other scenes. For example, in Roquentin's dream, Maurice Barrès tells a soldier with a hole in his forehead—presumably made by a bullet—that he should put a bouquet of violets into the hole. The soldier responds, "I'm going to stick them up your ass" (83). This soldier, Roquentin, and another soldier subsequently spank Barrès until he bleeds. In another scene, the Self-Taught Man is caught trying to touch young boys in the library. He is shamed and accused of pederasty in front of everyone. Roquentin equates his own experience of being taken from behind by existence with that of Lucienne: "Existence takes my thoughts from behind and gently expands them *from behind*; someone takes me from behind, they force me to think from behind, therefore to be something, behind me, breathing in light bubbles of existence . . . he says he would like to faint, he

of Western thought, see Robert Pogue Harrison's *Gardens: An Essay on the Human Condition* (Chicago: University of Chicago Press, 2008).

15. See the chapter "Metaphorics of the 'Mighty' Truth" in Blumenberg, *Paradigms*.

runs, he runs like a ferret, 'from behind' from behind *from behind*, little Lucienne assaulted from behind, violated by existence from behind" (139). The source of Roquentin's nausea is palpable but not visible, which leads him to try out various prepositions to locate it. Where is the phenomenon located? Is it behind? Beneath? Inside? Roquentin most commonly situates it beneath the surface of the visible. Like the root, it is hidden below. He describes the root, which triggers his crisis of being, as "below explanation" (129). The insufficiency of prepositions to describe what feels at times like a spatially determined apprehension is just one example of the failure of language throughout the narrative.

For Descartes, the root of the Tree of Philosophy was metaphysics. It is not surprising, then, that Sartre would choose the root as the site of the collapse of rational thought. The culminating scene of Sartre's book occurs in the Jardin public, the same park Roquentin traverses regularly in his digressions through Bouville and the same park where he witnessed "le petit drame" of the man in the cloak and the little girl. The illumination is provoked by the roots of a chestnut tree. It is probably a stretch to say that Sartre thought consciously of the mandrake legend when he chose the circumstances under which Roquentin has his illumination. But its elements are all there: spilled sperm from a criminal—in this case, a pedophile and perhaps a rapist and murderer—is necessary for the mandrake to grow. Perhaps it was spilled at that very bench at the foot of which Roquentin discovers the monstrous and terrifying botanical form. The tree's *Baumheit* or "treeness," as Heidegger would call it, asserts itself as the figure for Being itself: "So I was in the park just now. The roots of the chestnut tree were sunk in the ground just under my bench. I couldn't remember it was a root anymore. The words had vanished and with them the significance of things, their methods of use, and the feeble points of reference which men have traced on their surface. I was sitting, stooping forward, head bowed, alone in front of this black, knotty mass, entirely beastly, which frightened me. Then I had this vision" (171). The full passage illustrates the insufficiency of language, dissolving the connections between signifier, signified, and referent. Roquentin cannot recall that the root is called a root. Its label, a kind of veil that prevents one from seeing it for what it is, temporarily falls away unexpectedly as he contemplates this botanical form. The "feeble points of reference" that humans use to put objects in relation to themselves are such a given that when they dematerialize, what is left is too blunt and unbearable. Without the feeble points of reference, Roquentin is forced into a position of contemplating the verb "to be" and to acknowledge that being is the undifferentiated quality possessed by every object in the universe. His vision of being is not one of neutral, cold geometries. The mounds of being he suddenly perceives are soft, disordered masses, naked and obscene. They provoke *écœurement*, a nauseating disgust. Sartre chooses to describe existence in terms of abject forms because abjection is the most imposing experience a subject can encounter. Through its relentless inflections, the abject refuses to be ignored. It forces the beholder to contemplate what he would like to deny: death, filth, sexual taboo, and other unsettling phenomena. Had Sartre chosen a cold, geometrical depiction of what exists, Roquentin could have perhaps remained neutral or forgotten it quickly after having noticed it. Instead, he perceives it as an assault on his senses. He sees what others cannot, and is thus all the more plagued by the menace that is invisible to everyone else. He suddenly places himself in the category of all that is "de trop," in excess. He is superabundance in a human form. Like every object, he is too much, an irritating disruption of nothingness. Discovering that he shares the same ontological category as the revolting root, as the mud of Bouville, as someone's ugly hands striated with veins, puts him beside himself.

Other philosophers took inspiration from Sartre's jolting root scene for their own thought and writing. For example, Jacques Derrida read *Nausea* with "ecstatic bedazzlement."[16] Gaston Bachelard, who dedicates a section of his reflections on roots to Sartre's novel, describes Roquentin's root as "a sort of root that has lost its tree."[17] He notes further that Sartre refuses an ascensional depiction of the tree. He represses the tendency toward verticality, as delineated in chapter 3 of this study. Without verticality, transcendence is impossible.[18] For this reason in part, Sartre's narrative does not follow the transcendent arc common to mimetic realist fiction, particularly European fiction since the nineteenth century. In a chapter entitled "Roquentin's Nightmare," Robert Harrison argues that what disturbs the protagonist is the root's "begottenness," which "aggravates the opacity of existence."[19] Harrison's term "begottenness" is associated most often with procreation and the physical

16. "That I still admire and that I remember having read in a certain ecstatic bedazzlement at seventeen, in Algiers, in philosophy class, sitting on a bench in Laferrière square, sometimes raising my eyes toward the roots, the bushes of flowers or the luxuriant plants, as if to verify the too-much of existence, but also with intense movements of 'literary' identification: how to write like that and, above all, not like that?" Derrida, *Negotiations*, 264.

17. "Il s'agit d'une sorte de racine qui a perdu son arbre." Bachelard, *Terre*, 300.

18. "The refusal of the ascensional image—which is the most standard in the complete imagining of the tree—is furthermore clearly formulated by Jean-Paul Sartre. . . . this undoubtedly suffices to designate the *repression* of the usual image, the verticalizing archetype." Ibid., 304.

19. Harrison, *Forests*, 146.

bearing of children; one cannot forget its omnipresence in most translations of the Bible into English, in which this father begat that child in a seemingly infinite chain. "To beget" finds its French equivalent in the verb *engendrer*, whose Latin root *genus* may mean origin, birth, race, stock, descent, or kind. When a human is conceived, bodily matter coalesces from the genetic contributions of the parents and from the nourishment received from the mother. But the baby has not sprung out of nowhere. It was not pure nothingness before. The child is made up of recycled material. The begottenness feared by Roquentin is not the kind involved in baby-making. It is the much more basic fact of being itself, antecedent to the reorganizing of existing material to make new entities. His own existence, the kind owed to his parents' copulation, is only a figure for the more imposing fact of Existence. The metaphorical opacity to which Harrison refers is a way to express that the meaning of being cannot be gotten at. For most, there is a permanent impediment to perceiving it at all, much less to interpreting it. So even if the root imposes itself on Roquentin as a suddenly seeable figure for Being, it remains ultimately indecipherable and meaningless, opaque in its fundamental insignificance. Harrison also notes that the encroaching forest that threatens the city is a *topos* in Western literature, and Sartre's depiction of the botanical world in *Nausea* participates readily in this apprehension. The forest—"in which the fearful signs swarm," as Derrida wrote—is where the unknown takes up residency.[20] Roquentin writes in his journal that "once the city is dead, the vegetation will cover it,"[21] an image that recalls the sentient vegetation responsible in part for the fall of the House of Usher in Edgar Allan Poe's frightful tale. This fear of vegetal encroachment is evident in contemporary everyday life in the West, from the manicuring of urban botanical spaces to the availability of a vast array of weed-killers. Even Agent Orange and other herbicides used in the Vietnam War and various armed conflicts between the urbanized West and the forested East equate fear

20. Derrida, *Writing*, 88.

21. I am afraid of cities. But you mustn't leave them. If you go too far you come up against the vegetation belt. Vegetation has crawled for miles toward the cities. It is waiting. Once the city is dead, the vegetation will cover it, will climb over the stones, grip them, search them, make them burst with its long black pincers; it will blind the holes and let its green paws hang over everything. You must stay in the cities as long as they are alive, you must never penetrate alone this great mass of hair waiting at the gates; you must let it undulate and crack all by itself. In the cities, if you know how to take care of yourself, and choose the times when all the beasts are sleeping in their holes and digesting, behind the heaps of organic debris, you rarely come across anything more than minerals, the least frightening of all existants. (208–9)

with the forest and what might lurk within it.[22] From these dire warnings in the protagonist's journal, Harrison concludes, "Roquentin is thus condemned to the city, for the city remains the ultimate fortress of any humanism whatever."[23] The urban space is the last foothold of humanity in its conflict with the natural world.

Sartre shares with Heidegger a discomfort with Descartes's clean and schematic depiction of philosophy as a tree. Roquentin's crisis must occur at the root of the tree because one objective of Sartre's narrative is to call into question all of the assumptions and *idées reçues* about metaphysics. In the well-known introduction to *What Is Metaphysics?* [*Was ist Metaphysik?*], Heidegger takes Descartes to task, challenging the completeness of the latter's "Tree of Philosophy" metaphor. Heidegger begins by citing Descartes: "Thus the whole of philosophy is like a tree: the roots are metaphysics, the trunk is physics, and the branches that issue from the trunk are all the other sciences." In this same letter to his editor Picot, Descartes defined metaphysics as that which "contains the principles of knowledge/consciousness [*connaissance*], including the explanation of the primary attributes of God, of the immateriality of our souls, and of all the clear and simple notions that are in us."[24] Heidegger tries to push the metaphor further by introducing ground into the equation, suggesting that the more pertinent question involves the nature of the soil that nourishes the metaphysical roots. "What is metaphysics, viewed from its ground?," he wonders.[25] Heidegger is not alone here in his efforts to reground philosophy. As I show toward the end of this chapter, phenomenologists like Husserl and Merleau-Ponty also resorted to metaphors of rootedness and groundedness to describe what they felt the philosophy of their time had failed to take into account: the body as site of consciousness and the importance of subjective experience. Heidegger uses the capitalized Being to get at that quality so invisible to the human, that quality that Roquentin could not perceive until the root imposed its Being upon him and made him reckon with it. Roquentin's revelation dramatizes the recognition of Being that Heidegger believes should act as the soil on which the tree of philosophy feeds itself. In this overwrought and almost comical application of the metaphor cited above, Heidegger

22. In a passage from Céline's *Voyage au bout de la nuit*, Bardamu narrates, "Trees were something else I distrusted, ever since I'd been ambushed. Behind every tree a dead man." Céline, *Journey*, 47.

23. Harrison, *Forests*, 147.

24. Descartes, *Lettre-préface*, 32.

25. Heidegger, "Way Back," 207.

formulates a concept that he pursued throughout his writing years. Being and experience are "worlded"; they cannot be extracted from their context—their ground—under any circumstances. As shown in chapter 4, his obsession with autochthony and rootedness took many forms, from the insistence that places leave their signatures on people and cultural production to the more abstract, nearly allegorical version we see here. Despite Heidegger's claim that Sartre misread his work, we see clear echoes between the two thinkers through their critique of Descartes and their use of the tree to stage this critique.[26]

To conclude the analysis of *Nausea*, it is important to note the appearance in Sartre's narrative of Maurice Barrès, the anti-Dreyfusard from Lorraine whose nationalist, rooted politics I evaluate in chapter 4. He appears three times. In one particularly striking scene, Roquentin narrates a dream he had, beginning with this comical sentence: "I spanked Maurice Barrès" (83). It is no coincidence that the figure of Barrès, author of *Les déracinés* and a staunch defender of French regionalist essentialism, is punished by Roquentin. Yves Ansel argues: "Maurice Barrès incarnates all of the values condemned by the narrator [Roquentin]: the veneration of race and of the elite, the principle of authority, the respect for order, the glorification of military courage . . . , faithfulness to the earth, etc."[27] So, while Sartre uses the root primarily as a figure for the dissolution of Western metaphysics, the identity politics that haunted (and continue to haunt) France in his day are also addressed through the punishment of this polemical figure who advocated the expatriation of those who were not rooted in French soil.

SARTRE'S AUTOBIOGRAPHICAL TREE

Sartre's choice of the root as the site of the collapse of rational thought is not arbitrary. Trees had special significance for him personally; to some extent, he feared them. In his autobiographical text *The Words* (*Les mots*),[28] Sartre

26. Heidegger wrote: "Sartre expresses the basic tenet of existentialism in this way: Existence precedes essence. . . . Sartre's key proposition about the priority of *existentia* over *essentia* does . . . justify using the name 'existentialism' as an appropriate title for a philosophy of this sort. But the basic tenet of 'existentialism' has nothing at all in common with the statement from *Being and Time*—apart from the fact that in *Being and Time* no statement about the relation of *essentia* and *existentia* can yet be expressed, since there it is still a question of preparing something precursory." Heidegger, *Pathmarks*, 250–51.

27. Ansel, "*La nausée*," 37.

28. It is perhaps worth noting that Sartre's original proposed title for his autobiography was *Jean sans terre* (literally, "John without land"), a reference to John, king of England (1166–1216),

recalls a story called "Wind in the Trees" ("Du vent dans les arbres") from *Le matin*, which had frightened him as a child. He summarizes the plot of this story: a sick woman points through the window toward a chestnut tree that is shaking despite the absence of wind, and then she falls dead instantaneously. It is suspected that she saw an escapee from an asylum in the tree, but the truth remains concealed. Sartre refused the ending offered by the writer of the story: "According to the people of the village, it was Death that shook the branches of the chestnut tree."[29] Sartre writes, "I was afraid of the water, afraid of crabs and trees. Afraid of books in particular. I cursed the fiends who filled their stories with such atrocious figures. Yet I imitated them" (150–51). In a passage of *Truth and Existence* (*Vérité et existence*), he constructs a scene in which a tree is mistaken for a human being to illustrate how the mind projects identities upon the as-yet unidentified object.[30] The eye creates a fiction in which it falsely mimes "the vision of the tree."[31] Clearly, the category of the arboreal is marked

---

whose nickname was "John Lackland," a name given to him by his father to reflect the fact that he would not inherit any significant amount of property. That Sartre would choose a landless and thus unrooted figure for the title of his own autobiography is significant. When Simone de Beauvoir asked him what this title meant, he replied, "*Sans terre* meant without inheritance, without possessions. It meant what I was." Sartre quoted in Beauvoir, *Adieux,* 213. In his journal, Sartre wrote, "I've tried to destroy plenty of old ideologies, but my concern was to build. I may have lacked 'roots,' but I've never lacked stability." Sartre, *War Diaries*, 175.

29. Sartre, *Words*, 151.

30. "I can be entirely mistaken, take a tree for a boundary marker, think in the night that there is 'someone there' when there is no one; at least the tree is there, the night is there. In short, there is always something that is in-itself and whose initial unveiling is contemporaneous with my own surging up: Being is self-evident." Sartre, *Truth*, 21.

31. Persuaded that *something* is a tree, I generate the tree on that *something,* just as Kant insists that to perceive a line is to draw it. This means that I mime the vision of the tree, I retain each element of the vision in an organization called *tree.* I create what it is. If the in-itself *allows itself to be seen* as a tree, it organizes itself within my view in such a way that it answers the questions that my eye asks of it, so that my attempt to "see" this obscure mass "as branches" is crowned with success and suddenly a *form* constitutes itself that I can no longer undo.

Ibid., 22

In his notebooks, Sartre pursued a similar line of thought:

We may say, for instance, that perception of *that* tree is above all an existential phenomenon: to perceive the tree, for consciousness, is to surpass the tree towards its own nothingness of tree. One must not, of course, see in the word "surpassing" [*dépassement*] any indication of an *act.* It is merely a mode of existing. Consciousness exists for-itself beyond that tree as what *is not* that tree; the nihilating connection between

for Sartre by the limits of knowing; it can be confused for something else, hide something, or be "misread." The scenes of uneasiness provoked by trees are signatures of the occasional Gothic tendency of Sartre's existentialism.

The tree was for him a site of anxiety but also a catalyst for trained attention and contemplation. Sartre recalls a piece of advice from his grandfather: " 'Ah!' my grandfather would say, 'it's not enough to have eyes. You must learn to use them. Do you know what Flaubert did when de Maupassant was a little boy? He sat him down in front of a tree and gave him two hours to describe it.' I therefore learned to see."[32] But in another text about Ponge's "Le mimosa," Sartre argues that the exercise is absurd, that when the writer looks for something in a tree, whatever he happens to find, he put there himself. His conclusions, drawn ostensibly from the tree, are in reality emanations from his own psyche.[33] Nonetheless, as a young man, Sartre implemented this practice in his own life.[34] The tree is an ideal locus of contemplation for several reasons. It is stationary, which means it has no possibility of escaping our gaze, but its branches and leaves can stir in the wind. This combination of stillness and movement in one entity allows for a productive consideration of both states concurrently. The tree appeals to several of our senses simultaneously: we can see it, we can hear its leaves rustling, we can smell its sap. Its leaves are samples of the transformative power of nature: the plant converts sunlight into color and energy. In its yearly life cycle, visible in the changing palette of the leaves, we see a metaphor for human life and a living sign for the progression of the seasons. Part of the tree remains unseen, which elicits thoughts of invisible phenomena and their effects. The tree is alive, as we are, but cannot look back

---

> reflection and the reflected ensures that consciousness can be for itself only by reflecting itself as being, precisely, nothingness of the world *where* there is that tree. Which means it is non-thetic consciousness of itself as thetic consciousness *of* that tree; the tree is the transcendent theme of its nihilation.

Sartre, *War Diaries*, 179–80

32. Sartre, *Words*, 159.

33. "On rapporte que Flaubert disait à Maupassant, 'Mets-toi devant un arbre et décris-le.' Le conseil, s'il fut donné, est absurde. L'observateur peut prendre des mesures—et c'est tout. La chose lui refusera toujours son sens—et son être. Ponge regarde sans doute le mimosa; il le regarde attentivement et longtemps. Mais il sait déjà ce qu'il y cherche." Sartre, *L'homme*, 11.

34. "I began, in the Luxembourg [Gardens], by focusing my attention on a bright simulacrum of a plane-tree. I did not observe it. Quite the contrary: I trusted to the void, I waited. A moment later, its true foliage would suddenly appear in the form of a simple adjective or, at times, of a whole proposition: I had enriched the universe with quivering greenery." Sartre, *Words*, 182–83.

at us as we look upon it. This calls into question the ranking of forms of life: can a living thing be a subject if it doesn't have eyes? Looking at a particular tree recalls all of the mythological botanical forms from one's collective cultural heritage (the burning bush, Judas's suicide tree, the woods in which the Grimms' children are lost, the Tree of Life, Jack and the Beanstalk, Ovid's plant-human hybrids, Newton's apple tree). And because trees are found nearly everywhere, a thinker has easy access to a rich contemplative source at any time. In a well-known passage from *I and Thou*, Martin Buber articulates the rich potential of contemplating trees.[35] He enumerates the perspectives from which one may consider this botanical entity, foreign to ourselves as subjects. Husserl, too, in a section called "Noesis and Noema" in his *Ideas*, uses the tree as a point of departure for reflections on perception.[36]

The tree held another significance for Sartre. He took interest in this botanical life form because, according to Simone de Beauvoir, "the tree's pointless proliferation symbolized 'contingency.'"[37] Yves Ansel argues the following regarding the tree, and more specifically, the root, as the site of Roquentin's crisis: "That Sartre chooses precisely the tree to 'indicate' contingency does not originate in an unmotivated act: by means of the root, Roquentin interrogates and lays bare the very *grounds* of Western knowledge (and hence the emblematic value of the passage)."[38] In a letter to Beauvoir, Sartre describes in detail his contemplation of a tree, considering it a cognitive exercise. He claims it was in Le Havre that he "first understood the meaning of a tree."[39] It is worth noting that Sartre describes this contemplated tree—a chestnut tree, no less—as a victim upon which he exhausts his "arsenal" of comparisons, whereas in *Nausea*, Roquentin could be seen as a victim of the tree, which forces him to see what he doesn't want to. The imagination of the one who looks upon the tree

35. "I contemplate a tree . . . Throughout all of this the tree remains my object and has its place and its time span, its kind and condition. But it can also happen, if will and grace are joined, that as I contemplate the tree I am drawn into a relation, and the tree ceases to be an It. The power of exclusiveness has seized me." Buber, *I and Thou*, 57–58.

36. "Let us suppose that we are looking with pleasure in a garden at a blossoming apple tree. . . . The *tree plain and simple*, the thing in nature, is as different as it can be from this *perceived tree as such*. . . . The tree plain and simple can burn away, resolve itself into its chemical elements, and so forth. But the meaning . . . cannot burn away; it has no chemical elements, no forces, no real properties." Husserl, *Ideas,* 258, 260–61.

37. Beauvoir, *The Prime of Life*, 40–41. She claims that he wrote a poem called "L'arbre" ["The Tree"], which was subsequently lost.

38. Ansel, "*La nausée*," 54.

39. Beauvoir, *Prime of Life*, 89.

has the power to transform it into what it is not, to see behind it or around it or beneath it or to project upon it imagery that it does not yield readily on its own.

Roquentin's contemplation of the root leads him to reflect on what he calls the "fundamental absurdity" of existence. That which is absurd is out of tune, out of joint with its context. Language fails to make sense of the fact of the root, as do logic and explanation as well. Its presence is simultaneously undeniable and inexplicable. All *being* things (that is, all things) partake in this absurdity by their very nature. "This root . . . existed in such a way that I could not explain it," writes Roquentin.[40] It remains unnameable, a quality that Samuel Beckett took up in his text *The Unnamable* (1953). Absurdity was an essential motif for Beckett, who placed a single tree on the stage of his famous play *Waiting for Godot* (1952) to disorient his characters, to give a false point of reference in a universe that offers no bearings. This tree is symbolically rootless.[41] Despite the fact that it stands erect as though rooted in the ground, it is disconnected from any system. Everything in the blank world of the characters is incongruous. The first words readers encounter in Beckett's *Waiting for Godot* are these: "A country road. A tree. Evening," giving the semblance of a standard, contextualized dramatic setting. But the tree is a *fausse piste*, the wrong track toward understanding anything. When they first encounter it, Vladimir and Estragon try to use the tree as a point of orientation, because it was decided that Godot would meet them there. They are unsure whether it is the right tree, or whether it can even properly be called a tree at all. It shows a change of seasons, which gives an illusion of passing time or some kind of progression, but the stage is ultimately a closed circuit. Their world lacks signifying substance. As Michael Worton has argued, "The tree in *Godot* is a marvelous example of how Beckett refuses to allow concrete images to become (mere) symbols. . . . The tree thus means so *much* that it can have no single meaning."[42] To fill the hours of their waiting for a man who never arrives, they decide to do some exercises, then to imitate the bearing of a tree with erect posture and outstretched arms "for the balance."[43] The exercise is futile: as the stage directions indicate, both Vladimir and Estragon stagger in their attempts to hold still like the tree. They are rootless, stumbling for lack

40. Sartre, *Nausea*, 174.

41. In a notebook from 1940, Sartre calls himself an "abstract, rootless individual" (Sartre, *War Diaries*, 293) cut off from the central institutions and ideologies of his period (capitalism, parliamentarism, centralization, bureaucracy).

42. Worton, "*Waiting for Godot*," 80–81.

43. "Let's just do the tree, for the balance." Beckett, *Waiting for Godot*, 49.

of grounding. Toward the end of the second act, Estragon humbly proposes again that he and Vladimir hang themselves in the tree, a proposition already suggested in act 1. However, they abandon the plan and remain embedded in a contextless space in which real action is impossible and meaning is unattainable. It is not surprising that Ionesco defined absurdity as follows, using the root metaphor: "Absurd is that which is devoid of purpose. . . . Cut off from his religious, metaphysical, and transcendental roots, man is lost; all his actions become senseless, absurd, useless."[44] As Vladimir had concluded earlier in the play, "Decidedly this tree will not have been of the slightest use to us."[45] The absurdity of Beckett's staged world resembles the absurdity uncovered by Roquentin.

Nausea is not the only place where the root plays a pivotal role in Sartre's thought. His *Being and Nothingness* (1943), which is seen by many as the non-novelistic analog of *Nausea*, exposes many of the novel's themes using the language of philosophy. Perhaps unsurprisingly, the root metaphor appears here as well. In a section entitled "Doing and Having: Possession," Sartre juxtaposes the act of sliding with a state of rootedness. He compares consciousness to the way snow organizes itself to hold up the skis of a skier who glides down the slope, providing stable support during contact and then falling back away into a soft, yielding mass. Sartre considers the relationship of the skier to the white plain as a kind of possession. This sliding is "the opposite of taking root."[46] The kind of rootedness Sartre describes here requires a symbiosis

44. Ionesco quoted in Esslin, *Theater of the Absurd*, xix.

45. Beckett, *Waiting for Godot*, 48.

46. To slide is the opposite of taking root. The root is already half assimilated into the earth which nourishes it; it is a living concretion of the earth; it can utilize the earth only by making itself earth; that is, but submitting itself, in a sense, to the matter which it wishes to utilize. Sliding, on the contrary, realizes a material unity in depth without penetrating farther than the surface; it is like the dreaded master who does not need to insist nor to raise his voice in order to be obeyed. An admirable picture of power. From this comes that famous advice: "Slide, mortals, don't bear down!" This does not mean "Stay on the surface, don't go deeply into things," but on the contrary, "Realize syntheses in depth without compromising yourself."

Sartre, *Being and Nothingness*, 583–84

The aphorism cited by Sartre, "Glissez, mortels, n'appuyez pas," has been attributed to La Fontaine, Louis Sallentin, and Pierre-Charles Roy; the latter seems to be the definitive author of the line. The context of Roy's verse is

Sur un mince cristal l'hiver conduit leurs pas:
Le précipice est sous la glace;

between a subject and her environment, or that which she perceives without consciousness of it. If one invests the self too deeply in one's surroundings, one becomes inseparable from them. They are an extension of the self and, thus, virtually invisible. However, by skidding along the surface, letting things arise lightly in consciousness without bearing down upon them, the perceiving subject gets a sense of mastery over the encountered phenomena. They arrive and depart, delicate impressions left on a sentient and receptive surface.

PHENOMENOLOGY'S SEARCH FOR GROUND

Sartre stands alongside Husserl, Heidegger, Merleau-Ponty, and others as a pioneer in phenomenological thought. Phenomenology is sympathetic to the search for a philosophical ground from which thought should depart. So, while Deleuze and Guattari have argued that German philosophy is unique in its constant need to clear and consolidate a ground and to "lay foundations,"[47] the French were apparently as tempted by the phenomenological project to reground philosophy in a perceiving subject. This discipline takes interest in structures of consciousness as perceived and experienced from the first-person perspective. Phenomenology is less invested in reality than in the way a seeming reality manifests itself. For this reason, the perceiving subject and the phenomena that filter through the individual become the site of inquiry and the ground from which one can build a new understanding of consciousness.

What are the differences between groundedness and rootedness? Most obviously, in the fusion of earth and plant, groundedness focuses on the earth's role in this fusion while rootedness focuses on the plant's. From the perspective of groundedness, the earth plays the active role of holding and stabilizing

---

Telle est de vos plaisirs la légère surface.
Glissez, mortels, n'appuyez pas. Syers, *Poetry of Skating*, 43.

47. In the chapter "Geophilosophy" of *What Is Philosophy?*, they write:

> [German philosophy] wants to reconquer the Greek plane of immanence, the unknown earth that it now feels as its own *barbarism*, its own *anarchy* abandoned to the *nomads* since the disappearance of the Greeks. It must also constantly clear and consolidate this ground, that is to say, it must lay foundations. A mania for founding, for conquering, inspires this philosophy; what the Greeks possessed Autochthonously, German philosophy would have through conquest and foundation, so that it would make immanence immanent *to* something, to its own Act of philosophizing subjectivity (the cogito therefore takes on a different meaning since it conquers and lays down the ground).
>
> Deleuze and Guattari, *What Is Philosophy?*, 104–5

what is embedded in it, while from the perspective of rootedness, the plant clutches the soil, taking "responsibility" for its own uprightness. Furthermore, groundedness has other connotations, for example the notion of an electrical groundedness that allows currents to safely dissipate without risk of shock. To ground a child is to prevent his flight after disobedience, to preclude future disobediences. To be grounded is to be practical and level-headed, to have found a cosmic balance, to have established a solid base on which to build all activity and thought. To be rooted is to be anchored—for better or for worse—to a site. Those who are fearful are rooted to the spot. Rootedness allows one to tap into a collective history and to make use of the stores of accumulated cultural energy. While rootedness and groundedness have distinct connotations, they do have some shared meanings and are frequently used interchangeably.

Although proponents of phenomenology don't always use the root metaphor explicitly, their gesture is one of a turning toward the ground of experience. Many of these thinkers, however, do rely on the metaphorical potential of the root to express this shift in Western philosophy. I have shown how Heidegger appropriates and transforms Descartes's root metaphor to construct his own definition of metaphysics, but other phenomenologists rely on roots as well to describe what they believe philosophy should do. For example, Max Scheler argues uncompromisingly, "Every kind of cognition is rooted in experience."[48] In Husserl's introduction to the English translation of his *Ideas: General Introduction to Pure Phenomenology* [*Ideen zu einer reinen Phänomenologie und phänomenologischen Philosophie*], he writes that philosophy must become self-conscious of its own roots.[49] In addition to a specifically

48. Scheler, *Formalism*, 163.

49. Philosophy can take root only in radical reflexion upon the meaning and possibility of its own scheme. . . . There lies embedded in its meaning as philosophy a radicalism in the matter of foundations, an absolute freedom from all presuppositions, a securing for itself an absolute basis: the totality of presuppositions that can be "taken for granted." But that too must itself be first clarified through corresponding reflexions, and the absolutely binding quality of its requirements laid bare. That these reflexions become more and more interwoven as thought advances, and lead eventually to a whole science, to a science of Beginnings, a "first" philosophy; that all philosophical disciplines, the very foundations of all sciences whatsoever, spring from its matrix—all this must needs have remained implicit since the radicalism was lacking without which philosophy generally could not be, could not even make a start. The true philosophical beginning must have been irretrievably lost in beginning with presuppositions of a positive kind. Lacking as did the traditional schemes of philosophy the enthusiasm of a first beginning, they also lacked what is first and most important: a specifically philosophical groundwork

root-related vocabulary ("root," "radical," "embedded," "radicalism"), Husserl deploys several synonymous expressions for rootedness: "absolute basis," "foundation," "matrix," "start," "first beginning," "groundwork," "original," and "firmness of basis." The "original self-activity" he describes involves recouching consciousness in the body, which creates the preliminary conditions for its very possibility. Later, in his *Cartesian Meditations*, Husserl makes a similar assertion: "Philosophy . . . demands an elucidation by virtue of the *ultimate* and *most concrete* essential rootedness of any Objective world in transcendental subjectivity and thus make [*sic*] the world intelligible concretely: *as a constituted sense*."[50] He considers the objective certainty of positivist thought to be misleading and seeks instead an absolute point of departure, an unmitigated and stable position that takes into account the relevance of human experience. Philosophy itself must be taken to task. Without this first step—the clearing away of assumptions about how philosophical thought should operate—philosophers are doomed to remain in a faulty circuit that disregards the body and subjective experience, their task remaining thus incomplete. As David Abram describes eloquently: "The earth is thus, for Husserl, the secret depth of the life-world. . . . In his words, the earth is the encompassing 'ark of the world,' the common 'root basis' of all relative life worlds. Husserl's late insights into the importance of the earth for all human cognition were . . . to have profound implications for the subsequent unfolding of phenomenological philosophy."[51] Merleau-Ponty takes a similar position in his *Signs* (*Signes*, 1960) when he writes, "Reflection is no longer the passage from another order that absorbs that of actual things; it is first a more acute consciousness of our rootedness in them."[52] This attitude is manifest already in the moment of Merleau-Ponty's earliest public engagement. In his application statement for entry into the Collège de France, he insists on the "rootedness of mind *in its body* and *in its world*."[53] In his *Phenomenology of Perception* (*Phénoménologie*

---

acquired through original self-activity, and therewith that firmness of basis, that genuineness of root, which alone makes philosophy possible.

Husserl, *Ideas*, 27–28

50. Husserl, *Cartesian Meditations*, 137.

51. Abram, *Spell of the Sensuous*, 43.

52. Merleau-Ponty, *Signes*, 131.

53. "The mind that perceives is an embodied mind, and it is this rootedness of mind *in its body* and *in its world* that we've sought to bring back, against the doctrines that treat perception as the simple result of the action of exterior things on our bodies as well as against those doctrines that insist on the autonomy of awareness." Merleau-Ponty, "Un inédit," 402.

de la perception, 1945), he also describes perception in terms of the perceiving mind's grip (*la prise*) of the phenomena it encounters. The concept of grip, also implied in the etymology of *comprehend*, is suggestive of the relationship of the root to its environment, the way the plant clutches the earth for support. In English, the term *awareness* is less active than its French counterpart, *prise de conscience*, which implies a taking or gripping of consciousness. In his later thought, Merleau-Ponty developed the notion of flesh (*la chair*), which he viewed as a necessary addition to the four elements, as it is the only vital, living element.[54] Without it, the planet remains a dead one. As Sally Fischer has read this development in his thought, "Merleau-Ponty's concept of 'flesh' is an attempt to *root* his previous phenomenology in its fleshy (ontological) soil and to radicalize his critique of dualism by thinking anew the space between the subjectivist and objectivist alternatives."[55] The phenomenologists' metaphor of the body as site of the root of consciousness is a variation on the other kinds of human-plant syntheses I present throughout this book. Rather than rooting human consciousness in the ether, as Plato did, the phenomenologists embed it in the body and in lived experience. Some read this groundedness as a figurative "return to the earth," a new giving of attention to terrestrial concerns, including the well-being of the planet itself. In Charles Brown and Ted Toadvine's collection *Eco-Phenomenology: Back to the Earth Itself*, they argue that "phenomenology, as a contemporary method in philosophy, is particularly well suited to working through some of the dilemmas that have faced environmental ethicists and philosophers of nature."[56] They take the phenomenologists' vocabulary of a "return to the earth" as sympathetic with the goals of the ecological movement.

In an essay called "In Search of the Ground," the phenomenologist Eric Voegelin argues that a search for ground is common to all civilizations. The metaphysical form of this problem resides in the question, "Why is there something rather than nothing?" He recalls that the etymological root of the word "etiology"—the investigation of the causes of something, often, but not exclusively, a medical term—is *aition*, "ground" in Greek, which makes etiology "the quest for the ground." He shows how etiological endeavors that seek

54. "The flesh is not matter, is not mind, is not substance. To designate it, we should need the old term 'element,' in the sense it was used to speak of water, air, earth, and fire, that is, in the sense of a *general thing*, midway between spatio-temporal individual and the idea, a sort of incarnate principle that brings a style of being wherever there is a fragment of being. The flesh is in this sense an 'element' of Being." Merleau-Ponty, *Visible*, 139.

55. Fischer, "Social Ecology," 205.

56. Brown and Toadvine, *Eco-Phenomenology*, xi.

original causes and ultimate purposes have changed in nature through the centuries, beginning with foundational Greek thought and proceeding through the new derivatives in contemporary thought. From Voegelin's analysis, we begin to perceive in a wide range of disciplines (genealogy, medicine, etymology, myth, history) the relentless effort to pinpoint first causes or originary situations by which all futures are framed and determined. He writes: "It all comes back to the question: what is that ultimate purpose toward which we are rationally oriented? This leads us to the question of the nature of man, and to the answer that his nature differentially, as against all other creatures, is openness toward the ground. *That* is reason: openness toward the ground."[57] His very definition of myth describes it as a grounded phenomenon: "*Myth* can be defined . . . as imputation to other intracosmic things of a ground. It is myth when you tell a story of intracosmic ground" (233). He argues compellingly that the human tries to "misplace" the ground, seeking determinant original causes—everything from libido to race—that justify all human action (wars, sexual deviance, economic collapse). He argues that most of these root causes have simply been "rebranded" and recycled over the centuries, claiming: "We can observe, over the last two hundred years, that every possible locale where one could misplace the ground has been exhausted" (236). Voegelin is right to emphasize the centrality of the notion of groundedness—and, by extension, rootedness—in the twentieth century; few if any thinkers could exempt themselves from taking on this concept in that turbulent period. For Husserl, Heidegger, Merleau-Ponty, Voegelin, and other phenomenologists, what is incarnate or embodied seems to be synonymous with what is grounded or "worlded."[58] The conditions of being of the earth and of being in the flesh are imbricated notions. If the human, made of *humus*, did in fact spring figuratively from the earth—let us not forget that Adam was made of clay—the phenomenologists merely seek to remind us of the relationship between consciousness and the world that both created and continues to stimulate it.

As I have shown, Sartre's application of the root metaphor is perhaps the most multifaceted of any of its uses in the canon of twentieth-century French thought. He joins the other phenomenologists in his deployment of it but also makes it very much his own. While in the most famous example, he locates the site of a crisis in Western metaphysics at the root of a chestnut tree, the metaphor works very differently in his other uses of the motif. In *Being and*

57. Voegelin, "In Search," 232.

58. Metaphors of rootedness or groundedness abound in the work of phenomenologists like Edith Stein (a student of Husserl), Nicolai Hartmann, Jean-Luc Marion, and Paul Ricœur.

Nothingness, he depicts rootedness as the opposite of sliding, which he celebrates as a figure for consciousness that manifests itself freely and lightly, like the skier's grand gesture of appropriating a snow-covered mountain slope. The rootedness he describes as a counterexample of the slide involves a conflation of the self and the self's surroundings, which constitutes a loss of independence on the part of the conscious mind. So, for Sartre, the tree (a locus of contemplation and confusion) and, more specifically, the root (a locus of crisis, revelation, and radical assimilation) are both figural sites where humans negotiate their place in the world and their status as existing beings.

There are clear biographical indicators that the tree was significant for Sartre in his own life, but this accounts only in part for his repeated use of the root metaphor to convey various aspects of his existential and phenomenological thought. One could speculate that he was simply appropriating an old *topos*, inherited through philosophers such as Plato, Descartes, and Heidegger, positioning himself as an heir in their line of speculation and attempting to secure a spot in the pantheon of great thinkers by showing how he could take the metaphor and make it his own, adding a new development to philosophy, an old idea in a new key. I would suggest that Sartre's attraction to fiction plays a central role in his appropriation of the root metaphor. His fiction and nonfiction depictions of human encounters with trees are framed as narratives of meaningful and momentous convergence. For example, when Sartre narrates his contemplation of the plane tree in the Luxembourg Gardens or the chestnut tree in Le Havre, or when he places Roquentin before an oily, twisted root, these are not casual, everyday encounters; they are staged as scenes of revealed knowledge. Most people who look at trees do not perceive these occasions as momentous, but for Sartre, they are instances of revelation. Revelation, a powerful tool exploited by novelists and mystics alike in their narratives, involves the sudden presence of a previously absent knowledge. For Sartre and the phenomenologists, the root is the site, and in some cases the trigger, for a revelation of the unseen or the unacknowledged. Phenomenologists lamented the neglect of the root of consciousness, the blind spot of the philosophy of their time. For this reason, rerooting the human in what it could not or would not see was the objective of Sartre and his contemporaries. Because it is living but inert, because it is perceived as the hidden, primordial aspect of the tree, and because it has no defense against our beholding it, the root is an ideal site for the launching of a discussion about recontextualizing human consciousness in a world that seems to have expelled it. The phenomenologists used the image of a renewed human contact with the lost root to create a compelling case for their take on philosophy.

As I've shown in this chapter and will pursue in the following chapter, nearly every major philosophical breakthrough in twentieth-century Continental thought confronted the problem of rootedness explicitly. From phenomenology's subject-centered interrogations regarding ontological roots to the shift from a root-based to a rhizomatic notion of epistemology, from debates regarding the nature of radicality to identity-centered philosophical engagements on the nature of geographic, ethnic, or cultural belonging, twentieth-century European thought organized itself around tropes of rootedness, groundedness, implantation, transplantation, and eradication.

CHAPTER 6

Etymology and Essence: The Primeval Power of Word Roots

Etymologists, do not jump to conclusions! Do we not find two plants with quite separate roots sometimes mingling their foliage as one?

FRANCIS PONGE, *Méthodes*

As you have perhaps noticed, my study so far has made regular use of etymology in various ways. What is it I hope to show or accomplish by pointing out the occluded meanings found in the roots of words? While etymologies are often enlisted as incontrovertible proof of the latency of an idea or of the fact that we are mostly unaware of what we really *mean* when we speak or write, I can justify my "etymologizing method" by saying that for me, word roots are not particularly helpful as probative instruments but serve better as sites of the associative *potential* of words and concepts. If I invoke an etymology, it is not to suggest that I've discovered some primordial meaning hidden in the innermost kernel of a term but rather to show that forgotten roots sometimes give hints about why certain words evoke peripheral associations or to show that by recalling these roots, language can be poetically revitalized. My etymologizing gesture is only a secondhand gesture, the self-conscious reenactment of a practice that has for centuries served writers, philosophers, politicians, and theologians in Europe—and undoubtedly elsewhere—as a tool for convincing, justifying, legitimating, subverting, and playing.

In twentieth-century Continental Europe, a century and a place that witnessed two world wars, atomic proliferation, ideological clashes, decolonization, and increasingly asymmetrical terrorist maneuvers, many believed that a higher truth could be had if we could simply return to the root of things, approximating ourselves with an essence that had been covered by obstructive and debilitating layers. Language became once again a primary site for root-seeking. Like thoughts, language is imagined as a kind of plant that sprouts from a root and must be tended to. This explains why we call an anthology of literary excerpts a *florilegium*. The poet Edmond Jabès had the botanical aspects of language in mind when he titled his poem "V'herbe," a combination of "verb" and *herbe*, or grass.[1] From the retracing of etymological roots to the search for root metaphors, ground metaphors, or absolute metaphors, many aspects of language were interrogated in hopes of locating the primeval linguistic basis of thought and action. But as the century progressed, people seemed to lose faith in the noble root quest. Such a quest assumed first of all that the ancestors knew better and that language was a living bond with our forebears. This conviction, which requires a deferral and a deference to ancestral wisdom, insinuates that the present is necessarily less authentic than the past and that language moves from an original, concentrated essence to a watered-down, cloudy lexicon.[2] It is common practice to treat ancestors as infallible and infinitely wise. But a wealth of controverting evidence leads to the obvious question: "If they had it so right, how did it all go so wrong?" Furthermore, many doubted the assumption that if we could just find our lost roots and reestablish them in today's context, all problems would be solved. Such roots may have been functional in their original time and place, but their efficacy may have been context-specific and nontransferable to the complex intricacies of twentieth-century politics, philosophy, and social life. Finally, a loss of faith in the redemptive power of word roots was due to the misuse of etymology by many writers to justify their own agendas; mistaken or purely invented etymological reconstructions abounded during the period. With little restraint, these erroneous retracings were employed to reinforce political, theological, and ethical arguments and to lay claims to higher truth over one's enemies. Rediscovered etymologies—and even faked ones—were used often to legitimate a particular position and to undermine ideas that could not be etymologically supported.

The next section of this chapter investigates the obsession in the twentieth century with the retracing of word roots. Various thinkers turned to or away from etymology for a wide variety of reasons. Some of them placed all of their faith in the redemptive power of salvaged etymologies. Others considered it a

1. See Jabès, *Le livre des marges*, 109–12.

2. Derek Attridge writes, "Such a view of etymology implies the belief that the earlier a meaning the better, which must depend on a diagnosis of cultural decline . . . or a faith in a lost Golden Age of lexical purity and precision. . . . This view of etymology is one version of the widespread notion that words have *authentic meanings*, a notion instilled early in formal education and powerfully upheld by the ubiquity and status of the dictionary as a cultural institution." Attridge, *Peculiar Language*, 99–100.

playful space for punning and for refreshing language, a way to undo lexical petrification. Still others tried to subvert the reconstructionists' efforts by highlighting the misleading aspects of etymological devotion. Consistent among all of these various debates on etymology are the sustained themes of the subterranean, purity, source, and genealogy. As we have seen, these themes are part and parcel of the metaphor of rootedness, even in its linguistic iteration. The pure and hidden history of European culture finds a comfortable mythology in the very medium through which it is told: language. The various deployments of etymology in twentieth-century France and Germany presented in this chapter provide the clearest support of my claim that the more distant and unreliable the root, the easier it is to requisition it for ideological purposes.

THE ETYMOLOGICAL OBSESSION

> The fact is, man is an etymologizing animal. He abhors the vacuum of an unmeaning word. If it seems lifeless, he reads a new soul into it, and often, like an unskilful necromancer, spirits the wrong soul into the wrong body.
>
> ABRAM SMYTHE PALMER, *Folk-Etymology*

The textual production of the Italian philosopher and mystic Lanza del Vasto (1901–1981) evinces a desire for a return to something lost by humanity. From *Imaginary Etymologies* and *The Code of Things*, to *Pilgrimage to the Source* and *Principles and Precepts of the Return to the Obvious*, del Vasto's titles make obvious his feeling of urgency to return to an earlier state or to the fundament of all things in the universe. Del Vasto's curious biography, a mystically infused life trajectory comparable in many respects to that of Simone Weil, led him to desperately seek solutions for the human suffering he witnessed during the turbulent twentieth century.[3] In the introduction to del Vasto's *Les etymologies imaginaires* (*Imaginary Etymologies*), published in 1985 after his death in 1981, Pierre Souyris argued that del Vasto forced connections between etymologically unconnected words, simply because of homographic or homonymic similarity.[4] Souyris's reservations about these etymological reveries sums up

3. Born in Italy in 1901, del Vasto dedicated his life to peace activism and writing. He was a fervent Catholic who advocated an open dialogue between religions and who became a disciple of Ghandi. He traveled the world, organizing peaceful protests against the Algerian War and other armed conflicts. Del Vasto wrote many books, primarily in French.

4. "Like many philosophers, linguists, or poets (such as Valéry, Claudel, or del Vasto himself in *The Code of Things*), the author did not fail to be struck by how a 'pilgrimage to the source'

a very old debate that was rehearsed anew during the course of the twentieth century. Can etymologies be trusted? And if so, of what use are they to us?

The debate over etymology allowed for the formation of several opposing camps. Even defining the term is a contentious affair, as Mario Alinei has shown by comparing thirty-five different definitions of it.[5] For some, the attempt to pinpoint original meaning was approached as a highly scientific endeavor, with fringe linguists—fringe because etymology is generally marginalized as a pseudoscience even in the field of linguistics—trying to use as many resources as possible to declare with certainty a word's genesis. Naturally, the tensions between "scientific etymology" and "folk etymology" erupted throughout the century, as public sentiment over the legitimacy of the field remained unstable.[6] The more scientific-minded etymologists sought to wrest their discipline from the accusations lodged against it, namely that etymology is "anecdotal,"[7] "quaint,"[8] a "verbal pathology" (vii), a kind of "mischievous genius" (xxii), a "quasi-discipline," a set of "intellectual fables" or "scholarly fictions of connection,"[9] and a field that manifests "romantic amateurishness."[10] Some dismissed etymology altogether as a ridiculous fantasy with no bearing on the science of language, since it involves mainly anecdote and conjecture. Yakov Malkiel explains that etymology had little success as an "autonomous discipline, chiefly because it was associated with the reckless atomization of knowledge and, independently, because its pursuit was haunted, so rumour had it, by an excessive dosage of subjectivity and haphazardness."[11] Others

---

of words, a meditation on the mystery of their roots, can be fruitful and sometimes even deeply moving for thought. But what temptations one must avoid in this domain in which the craziest imagination can be given free reign, while so many homonyms and indeed homographs have nothing in common despite their appearance!" Souyris in del Vasto, *Etymologies*, 7–8.

5. See Alinei, "Thirty-Five Definitions."

6. Saussure was particularly distraught by what he called "popular etymology." In a comparison between popular etymology and analogy, he writes: "The only difference would be that analogical constructions are rational, whereas popular etymology proceeds randomly and merely produces howlers." Saussure, *Course*, 172. He continues his assault on popular etymology: "The degree of distortion thus creates no essential differences between words maltreated by popular etymology: what they all have in common is that they are purely and simply misunderstood forms which have been reinterpreted in terms of known forms" (173). His reasoning leads him to conclude that analogy is about remembering, while popular etymology is about forgetting (174).

7. Malkiel, *Etymology*, 43.

8. Palmer, *Folk-Etymology*, xvi.

9. Struever, "Fables of Power," 108.

10. Del Bello, *Forgotten Paths*, 27.

11. Malkiel, *Etymology*, 55.

acknowledged etymology's imprecision and lack of certainty, which, however, did not preclude an attempt to piece together a set of possible meanings using verifiable sources. They accepted their task as necessarily approximative but found it nonetheless significant. And for the most experimental thinkers, a relatively small cohort, etymology was seen as a creative practice, better suited for improvising new constellations of meaning than for locating old ones.[12]

Etymology is not the only kind of linguistic interrogation that seeks origin in the idiom. Language, perceived as an organic cord between present and past, is a site of hope for a recuperable purity buried by time. In Europe and the United States in the late twentieth century and early twenty-first century, an interest developed in tracing all metaphors and worldviews to their "ground metaphor" or "root metaphor," which, according to Stephen Pepper, is determined primarily by common sense;[13] in understanding the "mooring" or "grounding" of metaphor itself, as analyzed by M. Elaine Botha in her Christianity-focused book *Metaphor and Its Moorings: Studies in the Grounding of Metaphorical Meaning* (2007); or in identifying "absolute metaphors," which Hans Blumenberg identified as those metaphors that are the "*foundational elements* of philosophical language."[14] These metaphors are more than mere rhetorical flourishes that could be easily replaced with other figurative expressions or done away with altogether. Philosophy relies on these metaphors, Blumenberg

12. Yakov Malkiel writes, "Only under very special circumstances has etymology in the twentieth century been called upon to act not cognitively, but as an active force—as when certain poets fond of polysemy want the 'etymological meaning' to blend with other semantic shades of the word pressed into service." Ibid., 2.

13. Pepper, in his book *World Hypotheses: A Study in Evidence*, articulates the bold objective of his book this way: "Here I shall offer a hypothesis concerning the origin of world theories—a hypothesis which, if true, shows the connection of these theories with common sense, illumines the nature of these theories, renders them distinguishable from one another, and acts as an instrument of criticism for determining their relative adequacy." Pepper, *World Hypotheses*, 84. He defines what he calls the "root-metaphor method": "A man desiring to understand the world looks about for a clue to its comprehension. He pitches upon some area of common-sense fact and tries if he cannot understand other areas in terms of this one. This original area becomes then his basic analogy or root metaphor" (85).

14. "Metaphorology would here be a critical reflection charged with unmasking and counteracting the inauthenticity of figurative speech. But metaphors can also—hypothetically, for the time being—be *foundational elements* of philosophical language, 'translations' that resist being converted back into authenticity and logicality. If it could be shown that such translations, which would have to be called 'absolute metaphors' exist, then one of the essential tasks of conceptual history (in the thus expanded sense) would be to ascertain and analyze their conceptually irredeemable expressive function." Blumenberg, *Paradigms*, 3.

argues, and philosophers should be cognizant that meaningful contemplation cannot happen without them, since they create the conditions of possibility for certain ideas. Other metaphor studies focus on their experiential origins. In other words, just as there are some who believe that signs are not arbitrary, there are those who believe metaphors are not arbitrary but are instead rooted primarily in bodily experience. These studies attempt to access the primordial human through a piecing together of the cognitive jumps that manifest themselves in language over time. Already in 1936, Julien Pacotte, in his book *Le réseau arborescent, schème primordial de la pensée* (The arborescent network, primordial diagram of thought), enumerates the treelike structures common to widely varying systems, such as mathematics, cytology, biology, chemistry, and physics, and argues that the omnipresence of the arborescent form is not merely a question of convention; ramifying networks are "a universal aspect of intimate reality. It is the very foundation of formal thought."[15] Thus language, including its "evolving" history of metaphors or words, develops according to an arborescent logic because thought itself is already formally treelike. We feel a particular intimacy with the tree and its form because our language and thought naturally conform to its bifurcating pattern.

By no means was the twentieth century the first time etymology was considered a pursuit-worthy discipline. Perhaps in response to Barthes's declaration that "a whole history of etymological science, of etymological philosophy, from Cratylus to Proust's Brichot is yet to be written,"[16] several impressive histories of etymology have appeared since the 1980s. I will not summarize those histories but want to make clear that Europe has been heavily invested in the self-conscious tracing of its lexical roots at least since antiquity. Later examples, such as Isidore of Sevilla's famous seventh-century *Etymologiae*[17] (heavily relied upon by Aquinas, for example) and the ninth-century lexical encyclopedia *Etymologicum Genuinum* are among the best-known early European examples

15. "Les concepts qui procèdent nous mettent simplement en présence d'un objet formel inévitable, convenant originellement à la schématisation des ramifications qualitatives, qui sont un aspect universel de la réalité intime. Il s'agit ici du fondement même de la pensée formelle." Pacotte, *Le réseau arborescent*, 11.

16. Barthes, *Rustle*, 199. For an earlier, very compact history of etymology, see E. R. Curtius, "Etymology as a Category of Thought," in his *European Literature and the Latin Middle Ages* (Princeton, NJ: Princeton University Press, 2013), 495–500.

17. Nancy Struever writes that "Isidore of Seville's *Etymologiae* served as the program which schooled an elite's capacity to come to terms and exploit an alien body of inherited texts; it furnished a kind of medieval Ramism, a simple and easy method of dealing with books of legendary importance, in a time of cultural decadence." Struever, "Fables of Power," 109.

of a systematic effort at etymological reconstruction. Throughout the centuries, etymology was alternately considered a philological science of the simple retracing of word roots, an exercise in interpretation (*expositio*), an amateurish and anecdotal act of trivia, and a creative and experimental practice meant to expose new possible ways of using language.

In the twentieth century, thinkers and writers such as Valéry, Proust, Claudel, Heidegger, Agamben, and many others tended to look toward etymology as a kind of North Star with the power to orient the lost modern subject in a world that had forgotten its origins, now covered by centuries of historical debris.[18] The etymological obsession is a strong example of the uniquely human

18. Of Valéry, Malkiel writes,

> At the intersection of the semanticist's and the antiquarian's lines of thought one might place the procedure of a poet so learned as to have tendentially included, in his quest for elegant ambivalence, the favored key-words' etymological meaning as one of their many splendors, but neither so well-informed as to have selected a truly up-to-date guide to word origins, nor technically expert enough to have made forceful decisions of his own in the face of erudite controversy. Paul Valéry is rumored to have been one such poet, ever in the throes of etymological anguish; if this report is correct, his hidden etymologizing would be on a par with Flaubert's Carthaginian archeology, Rimbaud's evocation of tropical South America (known to him solely from hearsay and readings), and Tolstoy's logistics.
>
> Malkiel, *Essays*, 188–89

Proust's character Brichot, a pedantic rambler, is an avid believer in the power of etymology. Of the connection between etymology and genealogy in Proust, Sarah Tribout-Joseph writes:

> The practice of genealogy is not unrelated to that of etymology. Genealogies are, more often than not, drawn up to emphasize the importance of a present person by tracing them back to an illustrious predecessor. In Proust, the importance of this two-way movement is crystallized in the stained-glass window: in the right light, the one shines through the other. We are reminded of other lines of lineage, of Jesse Trees and also of the Bible with its long lists of who begat whom. The tree can "end" in disrepute, decadence, the undermining of the noble name. Paradoxically the desire for too much purity can also be undermining, with too much inbreeding resulting in senility (let us not forget that the duc and duchesse are cousins). Saint-Loup is the only specimen of the next generation of the Guermantes that we meet and he is killed in the war. Having throughout championed the discourse on virility, the Guermantes line is in danger.
>
> Tribout-Joseph, *Proust,* 139

In his *Art poétique,* Claudel writes in a note his translation of a quote from Plato's *Cratylus*: "Cratylus is right to say that natural names exist for things and that every man is not an artisan of names; only the one who considers which name is naturally suited to each thing and who knows how to reproduce the Idea of it in letters and syllables is such an artisan." Claudel,

sentiment of indebtedness to the past.[19] Feeling beholden to one's ancestors manifests itself in the search for the oldest etymological roots, which are perceived to contain some power that could impact the present. Having much more than a casual interest in the history of language, many dedicated etymologists believed that the kernels of words could contain a primordial key to understanding something hidden about the human.[20] By locating the *etymon*, we draw the ancestors near and implicitly ask them for advice on how to live and behave in the now. As Mario Alinei posits, "One could wonder whether the Indo-European short, monosyllabic form underlying English *I*, German *ich*, Latin *ego*, and Greek *egṓ*, etc., could not be seen as a 'primeval word,' the original motivation of which could correspond to the awakening of individual consciousness, at the dawn of mankind."[21] Such an idea clearly pursues the biblical notion of the Word as synonymous with Creation. The uttering of "I" equals the creation of consciousness. There are those, known as Cratylists or naturalists, who support the idea of an original, authentic meaning of a word that is not arbitrary but has some unique and nonnegotiable connection with the thing it names. The term *Cratylism* derives from Plato's *Cratylus* (fifth century BC), which stages a Socratic intervention in the debate between Hermogenes—who would today be called, like Saussure, a conventionalist, a believer in the arbitrariness of signs and in the powerful role of cultural convention in determining the names of things—and Cratylus, who argued for a nonarbitrary relationship between objects or people and their names. Although the definitive significance of the dialogue cannot be established with any certainty, Plato narrates a conversation that touches on questions with lasting relevance in the field of linguistics. This distinction between a naturalist and a conventionalist view of language largely determines a particular thinker's

L'art poétique, 172. George Steiner writes that "for Heidegger, the *figura etymologica*, the excavation of meaning from verbal roots and the history of words, is in the fullest sense an 'emergence,' a stepping into the light." Steiner, *Martin Heidegger*, 47.

19. For more on the indebtedness of human institutions to the legacies of the dead, see Harrison, *Dominion*.

20. Both published in the eighteenth century, two of the most important texts that explore the question of primordial language are Johann Gottfried Herder's "Essay on the Origin of Language" (1772) and Jean-Jacques Rousseau's "Essay on the Origin of Language" (1781). Two important texts whose objective was to locate the primordial origins of myths are Max Müller's *Essay on Comparative Mythology* (1858) and Ernst Cassirer's *Language and Myth* (*Sprache und Mythos—Ein Beitrag zum Problem der Götternamen*, 1925).

21. Alinei, "Thirty-Five Definitions," 19–20. Alinei does not support this hypothesis, but he considers it "viable" (20).

belief in the power of etymology. A naturalist would be much more seduced by the potential of etymological research to reveal the connections between things and their names. But if, as the conventionalist believes, these signs are arbitrary, why should etymology reveal anything aside from the haphazard pinning of names to phenomena? Saussure's insistence in the twentieth century on the arbitrariness of signs gave a boost to the Hermogenic position. If signs are arbitrary, then tracing words back to their roots is a fruitless exercise, first, because accurate reconstruction is impossible and, second, even if the origin of an utterance could be tracked with certainty, it would not get us any closer to truth.

The history of etymology in Europe can be considered through the standard polar optic of Hermogenism versus Cratylism, but the twentieth century added many fresh nuances to this old debate. For example, Gérard Genette, in his *Mimologics* (*Mimologiques: Voyage en Cratylie*, 1976), makes an important distinction between etymology and eponymy, a term, he argues, "that was missing for several centuries"[22] and thus impeded a true understanding of Plato's *Cratylus*. For him, the so-called etymologies in Plato's text were in reality eponymies, which are much more indirect in nature.[23] Genette argues that Plato "saw the trap. He also found the way out: the shift from indirect to direct motivation—that is, to sound symbolism." As Thaïs E. Morgan writes in an introduction to *Mimologics*, "Whereas etymology aims to trace words back to their historical origins according to the laws of filiation, eponymy allows for imaginative free play" (19). Derek Attridge pursues Genette's distinction, writing, "Word-play . . . is to etymology as synchrony is to diachrony."[24] Mario

22. Genette, *Mimologics*, 17.

23. "The eponymy of a person lies in the fact that he bears a significant name; the *eponymy of the name* lies in its value as a nickname, in the agreement between its designation and its signification, in its indirect motivation. By extension, we will say that eponymy as a 'science' (like toponymy) is the study of this type of motivation." Ibid., 16–17. He expounds,

> Eponymy serves to give a meaning to a name thought to be without one: that is to say, to find in it one or two hidden names, themselves hypothetically meaningful; or, in Proustian terms, to find the *names* hidden in the *words* . . . . The point of all this is clear: the problem with eponymical motivation is its *infinite* facility. It is an easy procedure, with a touch of complacency, of course . . . , but literally interminable. Every word is related to another, and so forth, until the inevitable return to the point of departure (since the lexicon itself is finite). (18)

24. Attridge, *Peculiar Language*, 109. Attridge compares etymology with the *calembour*, or paronomasia, the pun or play on words: "Two similar-sounding but distinct signifiers are brought together, and the surface relationship between them is invested with meaning through the inventiveness and rhetorical skill of the writer. If that meaning is in the form of a postulated

Alinei has proposed another way of thinking about etymology; he argues that it should be divided into two subfields, etymography and etymothesis. The first would involve "the description and careful reconstruction of the context in which a transparent word has undergone a semantic development," and the second would involve "speculations about the 'origin' of words" and "the reconstruction of the 'etymon' of opaque words."[25] Like Genette's etymology-eponymy distinction and Attridge's etymology–word-play distinction, Alinei seeks a terminology for the two primary ways an interest in word roots can be put into practice, either as a "scientific" method or as a more open-ended, speculative one.

The element of play introduced into the debate about the uses of etymology partly undoes the naturalist-conventionalist dichotomy by suggesting that etymologies, even false ones, have a potentially *creative* function. Here, poetics intervenes, and the etymologist-as-scientist is replaced by the etymologist-as-poet.[26] Roots of words are no longer chains that bind us to the past but transform instead into bifurcating strands that lead toward new semantic possibilities. The poets and writers who implement etymology for aesthetic purposes are thus exempt from the allegation of bad science. Paul Beekman Taylor has noted what he calls an "etymological style" in early Germanic literatures, which "draws attention to significations beneath designations—*les mots sous les mots*—or to a word's signifying source as well as to its signified sense."[27] Etymological poets like Mallarmé, Valéry, Claudel, Saint-John Perse,[28]

connection between present and past, what we have is etymology; if it is in the form of a postulated connection within the present, the result is word-play" (108).

25. Alinei, "Etymography," 47.

26. "That the etymologist is more of a poet than a scientist is by no means a new idea. Socrates, as he begins his etymological outpouring in the *Cratylus*, describes himself as one who is inspired and possessed." Attridge, *Peculiar Language*, 109. For more on the connection between poetry and etymology, see K. K. Ruthven, "The Poet as Etymologist," *Critical Quarterly* 11, no. 1 (1969): 9–37.

27. "In traditional, or native, Germanic literatures, there is a recognizable style which draws attention to verbal polysemy, a style that is etymological in the sense that it isolates word elements so that they manifest latent meaning. A meaning may be retrieved from an earlier, or original use, or may be discovered, or made, in context." Taylor, "Some Uses," 110.

28. According to Renée Ventresque, Jean Paulhan's text *La preuve par étymologie* (which is discussed in detail below) triggered Saint-John Perse's imagination regarding the possible uses of etymology for poetic inspiration, contrary to the text's polemical aim as an indictment of etymology's truth-revealing capacity. Ventresque goes on to argue that Saint-John Perse's etymological fantasies that unfold in the pages of *Amers* (1957) were inspired by both Paulhan's text and Wallace Fowlie's biography *Mallarmé*, which defines Claudel and Mallarmé as

Jean Cayrol,[29] Francis Ponge, and Velimir Khlebnikov and prose writers like Roger Caillois,[30] Michel Tournier,[31] and Hélène Cixous[32] use word roots as a loose remembering of past divergences in language. These divergences, forgotten or never known by contemporary users of the idiom, are recalled and recast to add dimensionality to the words on the flat page. As Ponge put it, etymology is the "poet's most necessary science."[33] For him, the roots of words "seem a little like the trunks of words, like the knot of being, their hardest, most solid, and most essential part."[34]

Recently, the British writer Robert Macfarlane has taken on the ambitious project of "rewilding" the English language by recovering a lost lexicon of place, collecting forgotten terms throughout the United Kingdom that describe landscapes, qualities of light or weather, and the subtle configurations

"cosmological poets" whose vocation was to discover the secret, originary unities of language. See Ventresque, "Bienfaits de l'étymologie."

29. In the introduction to *Les mots sont des demeures* (1952), Cayrol writes, "Words must be made inhabitable, they must be restored to their original splendor, their priceless innocence must be imposed." Cayrol, *Les mots*, 9.

30. According to his biographical profile on the website of the Académie française, of which he was a member from 1971 to his death in 1978, during the academy's discussions of changes to its dictionary, Caillois had the habit of proposing nonexistent words and substantiating them with very convincing faked etymologies—all in order to diminish the monotony of the meetings. See "Roger Caillois," www.academie-francaise.fr/les-immortels/roger-caillois, accessed January 31, 2014.

31. "To begin with, what *is* a monster? Etymology has a bit of a shock up its sleeve here: 'monster' comes from *monstrare*, 'to show.' A monster is something that is shown, pointed at, exhibited at fairs, and so on. And the more monstrous a creature is, the more it is to be exhibited." Tournier, *Ogre*, 3–4.

32. "I add. I open. I multiply geranium by its rich etymological hoard, I convoke the crane that conceals its name *geranos* in the meager bouquet, the screechy bird ger! ger! as if the flower in its tenacious structure were the flower of old age, I graft a gerontology." Cixous, *Hemlock*, 154. Cixous has used the word "root" in two of her titles: *Rootprints* (*Photos de racines*, 1994) and, more recently, *Le voyage de la racine Alechinsky* (2012), about the Belgian artist Pierre Alechinsky, who, Cixous argues, "enters art via the root" ("Alechinsky entre en art par la racine" [Cixous, *Voyage*, 22]).

33. "We may make fun of Littré, but we should make use of his dictionary. In addition to proper syntax, he does his best to sort out the etymology. What science is more necessary to the poet?" Ponge, *Nouveau recueil*, 15.

34. "Words must be used in their most certain, immutable meaning, the one that doesn't risk being lost in the future; thus, by preference, the words, or rather their *roots*, which seem to me a little like the trunks of the words, like the knot of being, their hardest, most solid, and most essential part." Ponge, *Pour un Malherbe*, 187.

of plant life. Of this project, titled *Landmarks* (2015), he writes, "I find lexicons to be more tropical jungle than tundra, gloriously ornate in their tendrilled outgrowths and complex root systems."[35] In Macfarlane's efforts to resuscitate language, one recognizes the attempt to recontextualize the human in nature. He laments the replacement of a vocabulary of the wild with a vocabulary of the technical, illustrative of humanity's divorce from nature. By recuperating the nature-centered lexicon of our forebears, we not only recover a lost genealogical connection; we also resituate the human in its original habitat. Macfarlane's objectives are also aesthetic in nature. By salvaging old words, vocabularies are expanded and expression is enriched; poetry can move away from the general and toward the specific; regionality will mean something again; and the monotony of a standardized, universal vocabulary can be undone with a new, beautified, detail-oriented idiom.

Thus etymologies and the renewal of a weakened lexicon can serve both ideological and aesthetic purposes. One finds a particularly strong example of the blurring of the line between etymology as poetic flourish and etymology as argument in Paul Claudel's etymological improvisations. Marie-François Guyard notes that in Claudel's *L'art poétique*, poetry and philosophy are bound together through etymology.[36] As I have shown, Claudel's biblical exegesis looks everywhere for evidence to support his particular reading of the Bible and his interpretations of Christian doctrine. Therefore, as is to be expected, etymology is just one among many tools he uses to support these interpretations. In his book *Contacts et circonstances* (1940), Claudel even goes so far as to suggest that the Bible predicted the alliance between Germany and the Soviet Union at the dawn of World War II (the Ribbentrop-Molotov Pact), basing this conviction on etymological proof.[37] Guyard diagnoses Claudel as having a chronic case of what he calls "etymologitis" (*l'étymologite*),[38] describing his etymologies as thorny (*scabreuses*) and stating unequivocally, "Claudel makes mistakes often and he does so in the name of an indefensible theory"

35. Macfarlane, "Word-Hoard."

36. "In the *Art poétique*, etymology does not only constitute a storehouse of curiosities; it is a method of exploring language and, through language, concepts; it is at the very source of philosophical reasoning." Guyard, "Claudel," 289.

37. See "La prophétie des oiseaux" in Claudel's *Contacts et circonstances.*

38. "Thus Paul Claudel presents all of the symptoms of a both chronic and acute *etymologitis*. Is it a congenital illness? It dates at least from youth." Guyard, "Claudel," 294. Guyard continues: "To these particular causes of Claudel's etymologitis, let us add a general cause: symbolism, both in the Mallarmean and Catholic sense, that makes him regard the universe and the Bible as encrypted documents" (296).

(286). And Claudel didn't stop at just words; like Mallarmé and Rimbaud before him, he saw letters as magical units as well.[39] In the end, Guyard concludes that Claudel was a bad philologist but a great poet, ultimately forgiving him for his overreaching etymological justifications.[40] Mobilizing any means available, including strained readings into cryptographic *tableaux* and etymologies, Claudel tries to find proof in both word and image of Christianity's truth, but ultimately he undermines the believability of these interpretations by pushing them to such a hyperbolic degree. His poetry, in its overemphatic Christian vehemence, can make the reader forget to read the verses as anything but relentless evangelism.

However, Claudel's poetic enlistment of etymology does highlight a signature metaphor attached to the discipline: that of the underground. As "ambassadors of a mute world," in Ponge's words, the poets are permitted if not encouraged to dig down deep into language.[41] As Mario Alinei argues, "etymology addresses the realm of opacity,"[42] that is, all that lies latent in the idiom. Depictions of etymological research tend to rely on a vocabulary of the subterranean. Robert Harrison, in his *Dominion of the Dead*, writes, "Dead etymons, latent meanings, and lateral connotations lie buried in the roots and phonemes of our living words, where they carry on an active afterlife."[43] Raimo Anttila compares the etymologist to an archaeologist,[44] echoing Charles de Brosses's term "archéologue universel."[45] Let us recall that Freud used a similar comparison—of psychoanalysis and archaeology—in the introduction to *Civilization and Its Discontents*, where he transforms Rome and its

39. "La *lettre*, ou plus précisément, la consonne, est une *attitude sonore* provoquée par l'idée génératrice, qu'elle *mime,* l'émotion, le mot." Claudel, *L'art poétique*, 171.

40. Guyard concludes: "Claudel n'est pas un philologue. . . . Mais il s'agit d'un grand poète . . . je veux dire que la connaissance de ses principes, même faux, en matière d'étymologie, peu contribuer utilement à l'intelligence de sa poésie." Guyard, "Claudel," 300.

41. "Poets should not get involved in their human relations but should sink down deep below. Furthermore, society takes good care of putting them there, and the love of things keeps them there. They are the ambassadors of the mute world. As such, they stammer, they murmur, they sink into the night of *logos*—until finally they find themselves at the level of the ROOTS, where things and formulations merge." Ponge, *Méthodes*, 205.

42. Alinei, "Thirty-Five Definitions," 16.

43. Harrison, *Dominion*, xi.

44. "The etymologist like any other archaeologist or historian may stumble on his subject accidentally." Anttila, *Historical*, 331.

45. In his *Traité de la formation méchanique des langues et des principes physiques de l'étymologie*, de Brosses gave the title "De l'archéologue ou nomenclature universelle réduite sous un petit nombre de racines" to chapter 16. See de Brosses, *Traité*, 456–96.

architectural history into a "psychical entity," the subconscious.[46] Jean Paulhan describes the etymologist's task as an imagined uncovering of the various layers that have concealed the word's meaning over time.[47] Railing against the etymological obsession, Jean Starobinski argues that the notion of the magic of an archaic word is nonsense,[48] and he is equally suspicious of the subterranean and primal aspects of the internal psychic life so dear to Freud, hidden in notions like the archetype, "the archaic word (or event)," and beneath the "luxuriant undergrowth of Fable" (334). Gaston Ferdière, a French poet and doctor with close ties to the surrealists, conceptualizes what he calls a lexical "speleopsychology," which entails an attempt at locating the subconscious origins of neologisms and, more specifically, *mots-valise*, or portmanteaus. He acknowledges the human temptation to dig for subterranean meanings in language.[49] As a solution, he proposes the practice of speleopsychology:

> If at times portmanteau words are obvious, there is no need to conjure up the magic wand of some benevolent or malicious fairy to demonstrate this; it should be left only to the "speleopsychologists" (if you'll allow me this neologism, which inscribes itself in the series of subterranean images encountered . . . from the pen of Michaux or Leiris) to seek the more or less distant "real origins," in the midst of accrued difficulties; for the subconscious has its rock slides, its chaos, its air currents, its crawlways, like the world of Martel or Casteret [the recognized founders of French speleology, the scientific study of caves]. (174)

In these depictions, language is spatialized and temporalized, composed of new strata that cover old ones. It is difficult to ignore the blatant phallic overtones of this linguistic penetration. If language is a composite of stratified

46. See Freud, *Civilization*, 16–17.

47. "Tracing the etymology of a word involves clearing away the inert layers that covered the first sound—the authentic sound—that formed it." Paulhan, *La preuve*, 21.

48. "I do not share the conviction that truth inheres in the archaic word, that it behooves us to dig deep with the new exegetical tools we have at our disposal in order to unearth permanent secrets of the 'human heart,' of 'human nature,' of 'being-in-the-world.'" Starobinski, "Inside," 333.

49. "The human mind likes to dig down into the caverns of vocabulary, to explore them, to expose itself to the greatest dangers; by leaning over the abysses, it catches sight of—and it does so *genuinely*—the explanations of its own ways of being and the answers to its own problems; this step does not satisfy it; it wants more, it leans further and *believes* it sees the answers to all its problems." Ferdière, "Vers une spéléopsychologie," 173.

layers to be penetrated by the etymologist, then the secret knowledge imparted through the process equals a kind of possession.

The theme of subterranean penetration leads effortlessly to another trope common in texts about etymology: that of purity. The traceable *etymon* represents a recuperable virginity. Pure language, unsoiled by time and the defilements of accumulated meanings, is the gift that awaits the committed etymologist. Foucault, for example, leans on this metaphor in *The Order of Things* when he writes, "To bring the origin of language back into the light of day means . . . to rediscover the primitive moment in which it was pure designation."[50] The narrative of ethnic purity, which dominated European politics of the twentieth century, and the much older Christian ideal of purity—associated with virginity, bodily cleanliness, rectitude, virtue, and the rinsing away of sins—worked mutually at the exclusion of the foreigner and the nonbeliever. The two groups could be conflated as figures of the impure. Nonmastery of the language, then, meant a kind of linguistic rootlessness and could be extrapolated as a sign of impurity and impiety.

The notion of purity is consistent with another common metaphor of origin in etymological descriptions: the source. By depicting word roots as emanating from a source, the qualities of water—the most purifying solution—could be enlisted to offer a built-in reason for a return to the wellspring of language, namely its qualities as a cleansing solution and a quencher of thirst. Liquefaction of the idiom makes language necessarily bound to the idea of life itself, given that water is the vital elixir. The portrayal of the origin of language as a source implies that it sprang up as a wonder of nature and will continue to offer itself forth spontaneously, in a cleansing surge. Of Renaissance etymologists, Nancy Struever writes, "Their return to the sources, *ad fontes*, was . . . a ritual of purification."[51] Derrida reflects on the problems of source, origin, and purity by posing the question this way: What is the source of the source?[52]

50. Foucault, *Order*, 104. See his chapter "Designation" for more on etymology and naming.

51. Struever, "Fables of Power," 110.

52. How can the source divide itself—sources that announce themselves as fertile already in the title—and thus separate itself from itself in order to relate itself to itself—which is, as a pure origin, self-unreference? And from the moment that the source opens its process, starts in on itself, and escape itself, is there a *first* metaphor of origin? An actual originary metaphor? A metaphor in which the source is lost less than in another? Or in which, losing itself more, it meets itself with more certainty? Is there in this procession—the vocabulary of Plotinus imposes itself here—a first metaphorical emanation of the One that is the source?

Derrida, "Les sources de Valéry," 573

By pushing the question back far enough, one realizes that the trouble with origin-seeking is that it always defers to an even earlier origin, a sustained gesture of regress that leads to nowhere but the imagined Beginning.

Derrida's reflections on the source lead to a final trope that is ubiquitous in the field of etymology: the metaphor of genealogy. In his well-known analysis "Nietzsche, Genealogy, History," Foucault, motivated to respond to Paul Rée's and Nietzsche's writings on the problem,[53] departs from the understanding of genealogy as the tracing back to some original source (of DNA, of moral or ethical first causes, etc.). Instead, Foucault argues that genealogy "opposes itself to the search 'origins'"[54] and that "a genealogy of values, morality, asceticism, and knowledge will never confuse itself with a quest for their 'origins'" (80). If genealogy is not a search for origins, then what is it? It is the "analysis of descent" (81); in other words, it thinks in the opposite direction; it is not a search for the past but an endeavor that uses the past as a starting point to follow subsequent lineages, including the unrealized lineages that just as well could have been. Emphasizing the accident, the error, and other deviations, Foucault nuances the meanings of various origin-related words used by Nietzsche, such as *Ursprung, Entstehung, Herkunft, Abkunft*, and *Geburt*, in order to show that descent, not origin, is the primary interest of genealogy. Etymology, often understood as the genealogical research of a word, could be recast through Foucault's analysis not as an attempt to locate an unfindable beginning but to analyze the concatenation of a word's lexical lives. Of most interest are the *relationships* between bifurcating strands. The original, primordial syllable is of little consequence. However, Foucault's emphasis on descent rather than origin has been historically ignored by most etymologists, particularly those who see high political stakes in their search for the primeval beginning.

An etymology is a story, a *Bildungsroman* of the word. Because the names of things are bound to life itself (the Word of God, Adam's engendering and naming of Eve, etc.), an etymology could be described as the biography of a word. Nancy Struever calls etymologies "stories describing *ur*-events of naming,"[55] and she calls etymologists "narrators" (112). Derek Attridge poses the question: "What are etymologies if not stories? What is the model for the history of the word if not the biographical narrative?"[56] Howard Bloch has

53. Foucault is most interested in Rée's *The Origin of Moral Sensations* (1877) and Nietzsche's *On the Genealogy of Morals* (1887).

54. Foucault, "Nietzsche, Genealogy, History," 140.

55. Struever, "Fables of Power," 108.

56. Attridge, *Peculiar Language*, 110.

noted the tendency since the Middle Ages toward a genealogical logic applied to linguistic fields like etymology. He argues compellingly that "a genealogically defined linguistic model informs not only the discipline of grammar . . . , but remains fundamental to an entire epistemological mode manifest in the discourse of history, theology, and Biblical exegesis."[57] In a close reading of the thirteenth-century *La queste del Saint Graal*, he notes the importance of the tree as the symbolic locus of genealogical transmission: "The metonymically linked trees imply a system of relations in which every part is, through the seamless web of a lost beginning (an original wholeness at the outset of time), embedded in a hierarchical and chronological series. . . . The decision to begin with the tree which preserves an attachment to origin, and which binds within a single nexus both semiology and kinship, is hardly innocent. For the *lignum* so enmeshed with the ideas of lineage and language also serves to define the terms and parameters for our discussion" (33). He goes on to show the manifold ways in which language (and thus meaning) and bloodlines are constantly imbricated in medieval texts, heraldry, and other material vestiges of the period. The connection was by no means a new one,[58] but it was revitalized in the Middle Ages and has remained a signature of European culture production ever since, although this link is largely overlooked today.

It is unsurprising that we find these familiar root tropes—the subterranean, purity, the source, genealogy—applied to etymological roots. The root cannot be "thought" without them, whether as a word root, an ancestral root, or the roots of a nation. As shown in chapter 1, particularly in Bachelard's and Jung's analyses of the root as a metaphor, the figure is charged with many subconscious associations, but these in particular seem fixed in the collective psyche. Language, like other immaterial forms of transmitted culture such as

57. Bloch, *Etymologies*, 28. It is clear that Claudel's exegetical and poetic methodology relies upon this connection between genealogy and etymology, which dates from the Middle Ages.

58. Bloch writes,

> The place of Babel as a dispersion of men and tongues is, of course, paramount in the linguistic mythology of the Christian West. What I am suggesting, however, is that the medieval reception of the Babel myth was itself part of a broader dynamic in which generation and signification are implicated in each other. Genealogy conceived along linguistic lines and language conceived along family lines represent two facets of a more general problematics of the sign prevalent in the thought of many of the most powerful intellectual figures from Augustine to the Renaissance. This connection is evident, for example, in the patristic fascination with the proximity of Adam's engendering of Eve and his naming of her.

Ibid., 35

concepts, is given a connective form that is both living and hidden. As a hobby for devoted word-lovers, etymology does not seem particularly threatening. However, Nancy Struever has pointed to potentially dangerous applications of the discipline. Using a Foucauldian framework, she shows the frequency and magnitude of ideological deployments of word histories.[59] Historically, etymology always went far beyond a leisurely pursuit of meanings for the sake of curiosity. The field, in varying degrees throughout history, has never been neutral or objective. "Etymology," writes Struever, "can become the vehicle for disguised dogmas, for a race or folk wisdom, and thus etymologies become fables of irrationalist power" (111). I will show how, in France and Germany throughout the twentieth century, etymology was used to lay claim to certain cultural lineages perceived as superior to all others. It goes without saying that the strategists of national myth-making relied heavily on etymological proof. The search for the roots of language, conceived in vitalist terms and reliant on the metaphor of purity, fits comfortably alongside the twentieth-century programs in eugenics, national myth-making, Christianizing, and assimilation of colonial subjects as crucial enterprises for the preservation of Europe.

GERMAN IDEOLOGICAL ETYMOLOGY

> We shall follow [etymologies] to wherever the kin of a word under discussion is, even if it has driven its roots beyond its own territory. For often the roots of a tree which is close to the line of the property have gone out under the neighbor's field.
>
> VARRO, *De lingua Latina*

Because etymology is such a loose field, without a firm consensus on its methodology or even its objectives, it has proved particularly vulnerable to

59. "Etymology was never solely a descriptivist tactic, simple antiquarianism, but always concerned with the issues of linguistic power. . . . Etymology as argument thus had great suppleness, a suspicious flexibility, and etymology as not very tightly disciplined power-play was from the beginning the scene of ideological confrontations and distrust." Struever, "Fables of Power," 110. Struever's primary argument is this: "I will argue that a history of etymological practice reveals a very sharp and suggestive antinomy: analysis of the discourse of 'classical' etymologizing uncovers a layer of egregious narrative realism: a policy which asserts a quintessentially 'historical' nature of linguistic force, and the quintessentially 'linear' nature of social identity. But 'modernist' etymologizing finds this narrative fundamentalism disturbing and rejects it. I maintain that the modernist subversion of narrative fundamentalism constitutes a cession of a domain of proper investigative interest, and is a subversion, or abdication, of intellectual range and disciplinary power" (108).

ideological appropriation and to strong influence by concurrent sociopolitical happenings or by theoretical shifts in other disciplines. Germany became the primary home of etymological research in the late nineteenth and early twentieth centuries. Yakov Malkiel notes that by 1900, etymology had achieved a certain measure of legitimacy throughout Europe but was "reputedly a quintessentially German subdiscipline."[60] Already in the late nineteenth century, Darwinism began to impact the field of linguistics in Germany, influencing specialists to define the development of language in terms of fitness and to organize it in families with genealogical offshoots to newer, fitter families. The German linguist August Schleicher, with his text *Darwinism Tested by the Science of Language* (*Die Darwinsche Theorie und die Sprachwissenschaft*, 1863), sought to show that the rules "which Darwin lays down with regard to the species of animals and plants, are equally applicable to the organisms of language."[61] It is not surprising that German linguists argued that their language was closest to Greek, "the language of philosophy," and that their term for the family of languages to which most European languages belonged is still referred to in Germany as "Indo-Germanic" (which indicates the southeastern-most and northwestern-most borders of this language family), rather than the term "Indo-European," whose use is widespread outside of Germany. Locating a possible *Urheimat* using etymological evidence was appealing to the proponents of National Socialism for obvious reasons. Gustaf Kossinna and Carl Schuchhardt, prominent specialists in prehistory, sought archaeological evidence of this *Urheimat* in the service of Nazi ideology. One can take the example of the swastika, chosen by the National Socialists as a symbol because of its "pure" Indo-Germanic history. Already in the 1870s, Heinrich Schliemmann offered evidence of this connection by calling the swastika an "exceedingly significant religious symbol of our remote ancestors,"[62] noting that it was "already regarded, thousands of years before Christ, as [a religious symbol] of the very greatest importance among the early progenitors of the Aryan races" (101–2). The symbol was thus reappropriated solely because of its supposed "rootedness" in Indo-Germanic culture. Efforts at retracing symbols to their roots mirrored the endeavors of etymologists. Malcolm Quinn has argued that "the malleable, compliant and mute prehistoric artefact" was the perfect place to argue for Aryan supremacy because it could be manipulated and customized

60. Malkiel, *Etymology*, 42.
61. Schleicher, *Darwinism*, 30.
62. Schliemann, *Troy*, 102.

at will.[63] Quinn's argument applies equally to etymology, a field that looks toward ultimately untraceable roots and often enlists its findings for ideological purposes. That "etymology is no innocent game"[64] is particularly clear in the years between the *Belle Epoque* and the years shortly following World War II. If the Germans dominated the field of etymology at the turn of the century, it was that same perceived "Germanness" of the field that became a liability during and after the war years.

In 1909, with Rudolf Meringer's founding in Heidelberg of the etymological journal *Wörter und Sachen*, a movement was established among Germanophone scholars that sought to study the connections between words and the material culture that influenced their coining. Proponents of the *Wörter und Sachen* movement attempted to bind meaning to something tangible rather than to speculative and often abstract etymologies disconnected from material culture.[65] This emphasis on "the thing" and its relation to language anticipates a vocabulary used by Heidegger and the phenomenologists for interrogating the nature of being and its representation through language. In fact, from Heidegger's claim that "the word makes the thing into a thing—it 'bethings' the

63. "Nineteenth-century Aryanism . . . was encumbered by the fact that the language and material remains of the proposed root-race were absent. This may explain why prehistory was so quickly enlisted to the cause: the malleable, compliant and mute prehistoric artefact, with none of the intractability, but all of the materiality of the written historical record, was ideally suited to the process that [Eric] Hobsbawm defines as 'identification.'" Quinn, *Swastika*, 34. Quinn refers here to the text *The Invention of Tradition* by Eric Hobsbawm and Terence Ranger (Cambridge: Cambridge University Press, 1983).

64. Guyard, "Claudel," 295.

65. In their introduction to the first issue, the editors write:

> After a period of wholesome restriction of linguistic studies to the investigation of phonetic changes, it seems the time has come to give the meanings of words, of "things," more attention again. By "things," we mean not only solid objects, but thoughts, ideas, and institutions as well, which find their linguistic expression in words. . . . The story of "things" is still not quite developed on all fronts, with large areas still obscure and with the material difficult to obtain; yet we take up etymological work if it keeps the objective of this unification [between things and words] in view, and will publish work dealing purely with the history of things, even if its use for the science of words belongs only to the future. . . . With changes in culture, words change their meaning. We demand that the explanation of changes in meaning not be sought in purely speculative ways but that they be dealt with in light of this fact.
>
> Meringer et al., "Vorwort," 1–2

thing,"[66] we might take a playful stab at creating a symmetrical formulation based on the tenets of the *Wörter und Sachen* movement: "The thing makes the word into a word—it 'bewords' the word." Naturally, after World War I, the reputation of the discipline of etymology suffered, given its perceived bond with German scholarship. Malkiel argues that World War II impacted the discipline even more, with etymological researchers purposely seeking "non-Germanic, preferably Latin" roots to words to avoid attributing lexical origins to the proto-Germans. Etymology was thus politicized to an extreme degree, and various groups vied for a claim on originary, and thus formative, lexicological contributions. Malkiel makes an obvious but important observation: "Such an atmosphere, permeated by politically flavoured pro and contra sentiments, seems unhealthy for the growth of impartial etymological research."[67]

Any conversation about the ideological uses of etymology in Germany would be incomplete without mentioning Heidegger, the "Pablo Picasso of language,"[68] whose etymological divagations have been the butt of many a joke.[69] Like Claudel, Heidegger was accused by many of having contracted a rather serious case of etymologitis. Karl Jaspers describes Heidegger's etymologism as the literalization of metaphors.[70] Using etymology as a "probe,"[71] his negotiation of the term *aletheia* in *Being and Time* is perhaps the most

66. Heidegger, *On the Way*, 151. "Das Wort be-dingt das Ding zum Ding." Heidegger, *Gesamtausgabe*, 220.

67. Malkiel, *Etymology*, 42.

68. "Aber gerade für die Sprachwissenschaft ist Martin Heidegger ein ebenso faszinierendes Phänomen wie Pablo Picasso für die Kunstwissenschaft." Wandruszka, "Etymologie," 860.

69. John Taylor argues that Heidegger is the butt of a joke in Raymond Queneau's *Saint Glinglin*: "As in all of Queneau's writings, the style is replete with wordplay, notably self-imposed lipogram: not one 'x' appears until the very last word, a constraint producing laughable, puzzling, and provocative spellings, especially when the author pokes fun at German philosopher Martin Heidegger and his famous analyses of etymology. 'Existence,' for example, becomes in turn 'aguesistence,' 'eggsistence,' 'eksistence,' 'ogresistence,' and 'hicksistence.'" Taylor, *Paths*, 51. Blanchot indicts Heidegger (and Hegel alike) for etymologism in *The Writing of the Disaster*. Alexander P. D. Mourelatos writes, "The *capricious use of etymology* in 'hermeneutic' interpretations of the pre-Socratics (by Heidegger and scholars who have been influenced by him) has given a bad name to etymological considerations as such." Mourelatos, *Route of Parmenides,* 197.

70. "Some, like Heidegger, would harness etymology to the task of tracing back through history the symbolism that is metaphor in language in order to recover literal root-meanings. These are to enable the modern philosopher to recapture a pre-historical, pre-symbolic level of human experience." Jaspers, *Tragedy*, 109.

71. Steiner, *Martin Heidegger*, 48.

cited example of this kind of experimentation. In an excellent summary of Heidegger's etymologizing practice, George Steiner shows how Heidegger's original use of etymology was to show in word roots "the greatest charge of initial and valid human perception," but in his later use of it, he begins to mine language to demonstrate that "the occlusion of this meaning has altered and damaged the destiny of Western thought" (7–8). Critiquing Heidegger's exceptionalization of Greek as a language unlike any other, as a language somehow more bound to original presence, Steiner continues: "In what way this assertion does no more than reproduce the allegories of Adamic speech and of Hebrew as we find them in Cabalistic and Pietist doctrines, and what conceivable means there could be of verifying Heidegger's claims, are legitimate and, indeed, urgent questions. What we want to know now, however, is just where this 'etymologizing realism' is taking us. The answer is: to yet further etymologies, which Heidegger's critics hold to be wildly arbitrary" (24–25). Steiner is right to identify the similarities between Heidegger's conception of language and Adamic speech. This account of linguistic beginnings allows him to nurse his attraction to the Mystery. Because others have already analyzed the numerous inaccuracies in Heidegger's etymologies, I will not do so here. Instead, I will only mention in passing that it is not surprising at all to find Heidegger turning to the primal origins of words to make his case about the uprootedness of modern life. Primordial word roots have the same appeal for him as the notion of autochthony, so central to his "Memorial Address" and to his earlier thinking about the relationship between people and their place of origin. As I have shown using the example of his literalization of the term *Bodenständigkeit*, Heidegger makes a habit of dusting off the particulate meanings that have collected over the centuries on words and ideas. His search for a primordial tongue is in keeping with his attempts to show the impact of the *Heimat* on cultural production. The authentic and the primal are for him coequal. Again, his etymological liberties attempt to make a case for the ancestral connection between Germany and Greece, and, as in his arguments about original autochthony, he relies on a mystical vocabulary of the primeval moment. Derek Attridge draws up a philosophical genealogy of how etymologies have been used by well-known thinkers, positing that "Vico and Heidegger may sound more like de Brosses and Tooke in their earnest search for 'authentic' meanings, whereas Plato's and Derrida's ironies play freely through their writing."[72] This distinction—between thinkers who themselves, in their use of etymology, truly believed in primordial authenticity and those who sought to play with the

72. Attridge, *Peculiar Language*, 125.

potential meanings of words—ultimately dissolves when their etymologies are recontextualized, reappropriated, and repurposed by later thinkers who cite these etymologies as truth.

PAULHAN'S ETYMOLOGICAL SKEPTICISM

The most focused criticism of etymologism to date was written by Jean Paulhan, an intellectual whose work was highly influential in his time but who is relatively marginal in the critical canon today. His 1953 book *La preuve par l'étymologie* (Proof by etymology), which has been described as an application *avant la lettre* of Derrida's *Of Grammatology* (1967),[73] was aimed at philosophers such as Alain (Emile Chartier), who, like many in his day, refused to acknowledge that very little difference separates an etymology from the *calembour*, or pun.[74] Paulhan launches a critique against this worst of superstitions, the belief that etymology provides some kind of deeper truth about language.[75]

73. "Paulhan gave us, in brief, a local instance of what might be called, before the letter, applied grammatology. That in 1953" (Mehlman, "Writing and Deference," 5); and "With the transcendental signified (or etymon) generated after the fact by a tension between signifiers, the problematic later to emerge as deconstruction was already broached" (8). Mehlman also describes deconstruction as "a forgetting of the perils engaged by Paulhan" (12). For a challenge to this genealogy, see Ann Smock and Jeffrey Mehlman's debate, "More on Writing and Deference," *Representations* 18 (Spring 1987): 158–64; page 14 of Ian Balfour, "'Difficult Reading': De Man's Itineraries," in *Responses: On Paul de Man's Wartime Journalism*, ed. Werner Hamacher, Neil H. Hertz, and Thomas Keenan (Lincoln: University of Nebraska Press, 1989), 6–14; and Syrotinski, "Domesticated Reading." Syrotinski writes, "If anything, what Derrida's and Paulhan's texts have in common is precisely a skepticism toward, or an ironization of, intellectual history structured as a continuous sequence of influences, which Mehlman's argument relies upon (including, most pertinently, the claim of a direct thread linking Paulhan to Derrida). In his reading of the connection between Paulhan and Derrida, Mehlman is not wholly off the mark, but very precisely half-way off the mark." Syrotinski, "Domesticated Reading," 91. For Paulhan's impact on intellectual circles in France, see Michael Syrotinski, *Defying Gravity: Jean Paulhan's Interventions in Twentieth-Century French Intellectual History* (Albany: State University of New York Press, 1998).

74. "Je vois bien qu'Alain prend appui sur une opinion fort courante et respectée. Elle n'en est pas plus respectable pour autant. Il suffit ici de mener une enquête précise—il suffisait aussi bien d'interroger les linguistes—pour observer que l'étymologie ne diffère en rien du calembour; et l'étymologiste ne découvre en général dans ses mots primitifs (dit-il) que ce qu'il a commencé par y mettre." Paulhan, *La preuve*, 101.

75. In a segment from his 1959–60 lectures, Jacques Lacan expresses similar reservations about the accuracy and validity of etymological research. His hesitations stem in particular from the very notion that language is "rooted." In a critique of Hans Sperber's 1912 text "Über den

He aims first at the assumption that words are motivated by some relationship to the physical world and that etymology allows us to trace language back to its more authentic, primordial root. Paulhan's book targets the search for "le sens authentique" (authentic meaning) through what he called the unveiling of language, a method consisting of a meticulous reconstruction of each word's lineage. Through a series of sarcastic examples, Paulhan pokes fun at thinkers and writers like Leibniz, Locke, Condillac, Comte, Alain, and Heidegger, those Cratylists who believe that words are not arbitrary. Even if words could be traced to their original meaning—a task he sees as impossible—his response to such success is: So what? What is to be done with an original meaning once it is found? He poses another pertinent question, "Must we believe that our ancestors were more intelligent than we?"[76] The truth of etymology is that of simple anecdote, an unsubstantiated story that may be amusing, strange, or fascinating, but which does not actually tell us anything about language or about culture. He compares etymology to the *calembour*, or pun, explaining that neither of these rhetorical forms keeps its promise as an explanation about the relationship between multiple words or concepts.[77] Etymology is not a

---

Einfluss sexueller Momente auf Entstehung und Entwicklung der Sprache" (On the influence of sexual factors on the origin and development of language), Lacan challenges the metaphor of linguistic radicality and comes to the conclusion that the "notion of root is highly tenuous":

> That words whose meaning was originally sexual spread out so as to overlay meanings that are very remote doesn't mean as a consequence that the whole field of meaning is overlaid in that way. That doesn't mean that all the language we use is in the end reducible to the key words it contains, words whose valorization is considerably facilitated by the fact that one accepts as proven what is, in fact, most questionable, namely, the notion of a root or a radical, and what in human language would be its constitutive link to sense. This emphasis placed on roots and radicals in languages making use of inflections raises particular problems that are far from being applicable to human language universally. What would be the case with Chinese, for example, where all the signifying units are monosyllabic? The notion of root is highly tenuous. In fact, what is involved is an illusion that is linked to the development of language, of the use of the language system, which can only seem very suspect to us.
>
> Lacan, *Ethics*, 167

Using this comparative approach, Lacan shows that this difference between Chinese and European languages refutes the notion that all languages began with elementary units that eventually clustered or burgeoned to form more complex units in time.

76. Paulhan, *La preuve*, 18.

77. Of *calembours* and etymologies, Paulhan writes: "D'où vient sans doute le caractère comique de telles phrases: c'est qu'elles ne tiennent pas la promesse qu'elles ont un instant paru nous faire, et ne nous offrent qu'une explication sitôt ruinée qu'ébauchée." Ibid., 66.

science but a rhetorical game that "justifies anything you want, without teaching us anything" (62). Paulhan's critique aligns with my larger argument in this chapter on word roots and with my argument about rootedness in general. Belief in rootedness of any form—linguistic, genetic, geographic, cultural—shows the particular weakness of our species, which possesses the gift and curse of remembering. A person with no memory of her yesterday or of the yesterdays of her culture would not feel the need to define herself or "locate" herself in terms of her ancestors. She would have no concept of her language as a lexical heap under which purer meanings hide. She would not feel inferior to her forebears, nor would she measure her own actions by their accomplishments. Certainly, we are all the products of accumulated culture, but why should we feel obliged to measure our own legacy in terms of what must certainly be a misperception of that of our forebears? Given that even our daily assessments of our own culture are always contentious, riddled with errors and misrepresentations, how could we possibly be certain to have understood the remote roots that we lean upon so heavily to justify contemporary actions? This challenge to etymological primacy was taken up by the poststructuralists, particularly throughout the 1970s and 1980s, and it continues to play a role in contemporary French thought.[78] For Jacques Derrida and other proponents of deconstruction, etymological play subverts the narrative that what is old is more true and that word origins can be authoritative sources for truth. Instead, he sought to show the inherent instability of language, with words at times meaning simultaneously one thing and the very opposite.[79]

78. Cécile Hanania dedicated an entire book to the relationship between Roland Barthes and etymology, for example. She writes in the introduction, "Throughout his critical development and regardless of which theoretical movements one associated with him or that he belonged to, Roland Barthes appealed constantly to etymological filiations or deductions—proven or imagined—in his analyses and reflections." Hanania, *Roland Barthes*, 16. In his use of the discipline, Barthes seemed to oscillate between etymologies used as a legitimate philological tools and as "etymologies constructed at the whim of imaginary and playful associations" (201–2).

79. Derrida's analysis of the *pharmakon* in his essay "Plato's Pharmacy" is perhaps the best-known example of this argument. In "The Double Session," he also mentions Karl Abel's "On the Antithetical Sense of Primary Words," of which Freud wrote the following assessment in a review of the work: "To the chance reading of a work by the philologist K. Abel I owe my first understanding of the strange tendency of the dream-work to disregard negation and to express contraries by identical means of representation." Freud, "Antithetical," 56. Freud continues: "In the agreement between that peculiarity of the dream-work mentioned at the beginning of this paper and this which philologists have discovered to be habitual in the oldest languages, we may see a confirmation of our supposition in regard to the regressive, archaic character of thought-expression in dreams. And we cannot dismiss this conjecture, which forces itself on

DERRIDA'S DERACINATION OF LANGUAGE

Derrida has been regarded by many as a sort of trickster figure, interested mostly in puns and language games rather than in a more legitimate and serious engagement with the texts he studies.[80] In analyzing Derridean applications of etymology, it is useful here to reiterate the distinction between word history and word play. As we have seen, Derek Attridge puts Vico and Heidegger in the camp of authenticity-seeking etymologizers, while he considers Plato and Derrida to be etymologizers who let their "ironies play freely through their writing."[81] Of deconstruction's linguistic play, Emily Apter writes: "Through diacritical invention, language was defamiliarized. Neologisms and syntactic intercessions broke up patterns of impacted, predictable meaning. The separation of prefixes and suffixes from verbal *racines* released lost or forgotten significations into the imagination."[82] Derrida was clearly on a mission. Because his texts resist interpretation and comment, I will limit my discussion of them to a survey of the instances in which rootedness surfaces as a pressing problem, and which, I argue, contain in a compressed form the entirety of his thought. An appraisal of his corpus makes it clear that the problems of origin, ground, root, foundation, source, and genealogy were perhaps his key intellectual preoccupations for the length of his writing life.[83] His scrutiny of the problem

---

us psychiatrists, that we should understand the language of dreams better and translate it more easily if we knew more about the development of language" (62).

80. Richard Rorty writes: "Derrida does not want to comprehend Hegel's books; he wants to play with Hegel. He doesn't want to write a book about the nature of language; he wants to play with the texts which other people have thought they were writing about language." Rorty, "Philosophy," 147. Nancy Struever argues, "Take Jacques Derrida: his etymologies are mostly jokes, and since the aim is subversion, it is difficult to say whether the plots are in good or bad faith." Struever, "Fables of Power," 118.

81. Attridge, *Peculiar Language*, 125.

82. Apter, *Continental Drift*, 8.

83. His philosophy thesis at the Ecole Normale Supérieure was entitled "Le problème de la genèse dans la philosophie de Husserl." His texts, such as *On the Name* and *Monolingualism of the Other; or, The Prosthesis of Origin*, have obvious investments in destabilizing linguistic foundations. *Archive Fever*, too, explores the relationship between thinkers of today and the texts of their forebears, a relationship necessarily fraught with apprehension. He also sought to critique others' positioning on various "original" concepts, for example in his engagement with Condillac's *Essai sur l'origine des connaissances humaines* in the book *Archéologie du frivole*, in his reading of Heidegger's essay "The Origin of the Work of Art" in his essay "Restitutions" (*The Truth in Painting*), and in his reading of Husserl's *Origin of Geometry* in *Voice and Phenomenon*.

of writing is an apt departure point, since it was this theme that dominated his first public interventions. He sought to show—first through theory, then through practice—how Western philosophy has taken for granted the role of writing, which assumes itself as a given, the medium through which philosophical thought is necessarily communicated. In *Of Grammatology*, Derrida posits that our "philosophical conceptuality" is "entirely commanded by a situation determined by the relationships between logos and writing."[84] He makes an effort to force philosophers to acknowledge that their arguments are largely metaphorical in nature and that this metaphorization is simultaneously self-justifying and self-veiling.[85] In a signature deconstructive move, Derrida uses metaphor—"sediment," "entangle in their roots"—to emphasize the same metaphorizing philosophy he critiques. His conclusion here—that "this abyss of metaphor will never cease to stratify itself"—is followed by a lifelong performance and an endless cycle of texts that seem to feed on themselves. In the context of pre–May 1968 Paris, it is easy to imagine the kind of resistance such ideas would encounter, particularly at his home institution, built on the very foundations he sought to topple. In a cautionary introduction to Derridean vocabulary, Simon Morgan Wortham emphasizes the difficulty in

84. Derrida, *Of Grammatology*, 81.

85. The appeal to the criteria of clarity and obscurity would suffice to confirm what we stated above: this entire philosophical delimitation of metaphor already lends itself to being constructed and working by "metaphors." How could a piece of knowledge or a language be properly clear or obscure? Now, all the concepts which have operated in the definition of metaphor always have an origin and an efficacity that are themselves "metaphorical," to use a word that this time, rigorously is no longer suitable to designate tropes that are as much defining as defined. If we went back to each term in the definition proposed by the *Poetics*, we could recognize in it the mark of a figure (*metaphora* or *epiphora* is also a movement of spatial translation; *eidos* is also a visible figure, a contour and a form, the space of an aspect or of a species; *genos* is also an affiliation, the base of a birth, of an origin, of a family, etc.). All that these tropes maintain and sediment in the entangling of their roots is apparent. However, the issue is not to take the function of the concept back to the etymology of the noun along a straight line. We have been attentive to the internal, systematic, and synchronic articulation of the Aristotelian concepts in order to avoid this etymologism. Nevertheless, none of their names being a conventional and arbitrary X, the historical or genealogical (let us not say etymological) tie of the signified concept to its signifier (to language) is not a reducible contingency. This implication of the defined in the definition, this abyss of metaphor will never cease to stratify itself, simultaneously widening and consolidating itself: the (artificial) light and (displaced) habitat of classical rhetoric.

Derrida, *Margins,* 252–53

approaching Derrida's philosophy as authoritative.[86] To consider it a ground upon which to make claims or to gain some broad understanding of Western civilization is a risk. Derrida defines himself in opposition to two figures—Heidegger and Husserl, the roots of his own philosophical patrilineage—to whom he feels indebted and against whom he simultaneously launches his insurgence. While they took as their task a regrounding of metaphysics, their "son," in mutinous fashion, sought instead to uproot everything, even philosophy itself. At times, Derrida is explicit in his critique of the root metaphor in all its forms and its hegemony over Western thought, writing that the root as a figure for origin is most often in truth "a concealment of the origin."[87] Launching a critique of the system as a whole rather than of a particular movement or school of thought, he uses the three heraldic texts published in 1967 to make clear that a such a critique requires a virtual barricade of text, a triptych, each piece of which echoes and sustains the others, offering to the public in one barrage something to talk about and a fresh vocabulary in which to do so.[88] It is fitting that the problem of origin is central in these debut publications; at the moment of his own initiation into public philosophy, he begins with the beginning. In *Of Grammatology*, Derrida poses the question, "What is a lineage in the order of discourse and text?," raising the implicit objection to a genealogical

86. The notion that this book gives a *grounding* in "Derrida" should be treated with considerable caution. As its readers will discover, that which *founds* or *institutes* always imposes itself, for Derrida, more or less violently, more or less justifiably, taking possession of its ground at the price of significant exclusions and contradictions. For Derrida, too, the metaphysical tradition which bases itself on the determination of being as presence is maintained on the strength of a repression of *différance* as the nevertheless groundless "origin" which, in fact, constitutes that tradition's very possibility. For Derrida, then, one must carefully consider the limits and implications of establishing one's grounds.

Wortham, *Derrida,* 1

87. Indeed, one must understand this *incompetence* of science which is also the incompetence of philosophy, the *closure* of the *epistémè*. Above all it does not invoke a return to a prescientific or infra-philosophic form of discourse. Quite the contrary. This common root, which is not a root but the concealment of the origin and which is not common because it does not amount to the same thing except with the unmonotonous insistence of difference, this unnameable movement of *difference-itself,* that I have nicknamed *trace, reserve,* or *difference,* should be called writing only within the *historical* closure, that is to say within the limits of science and philosophy.

Derrida, *Of Grammatology,* 93

88. These three texts, *Of Grammatology, Voice and Phenomenon*, and *Writing and Difference* were published by three separate publishers: Minuit, PUF, and Seuil, respectively.

depiction within a text or intertextually, but then he immediately forecloses the possibility of an answer with what he sees as an incontrovertible fact: "We know that the metaphor that would describe the genealogy of text correctly is still *forbidden*" (101). He continues, leaning once again—this time more heavily—on the metaphor of the root:

> If a text always gives itself a certain representation of its own roots, those roots live only by that representation, by never touching the soil, so to speak. Which undoubtedly destroys their *radical essence*, but not the necessity of their *racinating function*. To say that one always interweaves roots endlessly, bending them to send down roots among the roots, to pass through the same points again, to redouble old adherences, to circulate among their differences, to coil around themselves or to be enveloped one in the other, to say that a text is never anything but a *system of roots*, is undoubtedly to contradict at once the concept of system and the pattern of the root. But in order not to be pure appearance, this contradiction takes on the meaning of a contradiction, and receives its "illogicality," only through being thought within a finite configuration—the history of metaphysics—and caught within a root system which does not end there and which as yet has no name. (101–2)

The racinating function of a text and its role as the point of stability in a discourse, the ostensibly unchangeable and written truth that anchors generations who cannot communicate with one another face to face, has yet to be acknowledged, he argues. That this function exists, goes unnoticed, perpetuates certain ideas about the codependency of thought and writing, and transmits itself from generation to generation without anyone bothering to ask why is an aggravating problem for Derrida. Regarding any text, but particularly those that try to articulate a "genealogical representation," he insists on the necessary copresence of several periods inside it.[89] The invisible, self-involving strata of the text pretend to relieve us of the work of starting from the beginning with each new line of thought, but this work left undone means that the writer unwittingly allows the text to prohibit a critique of the Text as a concept. In *Writing and Difference*, which accentuates problems of origin, genealogy, and

89. "A text always has several epochs and reading must resign itself to that fact. And this genealogical self-representation is itself already the representation of a self-representation; what, for example, 'the French eighteenth century,' if such a thing existed, already constructed as its own source and its own presence." Derrida, *Of Grammatology*, 102.

rootedness, Derrida calls writing "the passion of the origin."[90] In addition to the rootedness of philosophy in *logos* and writing, the Derridean vocabulary—composed of terms like *trace*, *ghost*, *archi-writing*, *paleonym*, and *khôra*, all of which function frequently as synonyms for "root"[91]—calls attention to the innumerable ways the past imposes itself on the present. The image of cinders or ashes (*cendres*), to which Derrida dedicated an entire book,[92] makes its way from Foucault's *History of Madness* as the "charred root of meaning" ("la racine calcinée du sens")[93] into Derrida's "Cogito et histoire de la folie"[94] (an article about Foucault's text), into his *Circumfession*, and finally into his collaboration with Geoffrey Bennington called *Derrida*.[95] Simon Wortham writes, "For

90. "Writing, passion of the origin, must also be understood through the subjective genitive. It is the origin itself which is impassioned, passive, and past, in that it is written. Which means inscribed." Derrida, *Writing*, 373.

91. See in particular the chapter "The Voice that Keeps Silent" ("La voix qui garde le silence") in *Voice and Phenomenon*, where, in his critique of Husserl, Derrida makes claims such as "Temporalization is the root of a metaphor that can only be originary" and "The trace is unthinkable if we start from the simplicity of a present whose life would be interior to itself. The self of the living present is originarily a trace. The trace is not an attributable about which we could say that the self of the living present 'is originarily' the trace. It is necessary to think originary-being from the trace and not the trace from originary-being. This archi-writing is at work in the origin of sense." Derrida, *Voice*, 73. In the English edition, the translator, Leonard Lawlor, writes that *Voice and Phenomenon* "contains the 'germinal structure' of Derrida's entire thought" (xi). Of the final term, Derrida writes: "*Khôra* marks a place apart, the spacing which keeps a dissymmetrical relation to all that which, 'in herself,' beside or in addition to herself, seems to make a couple with her. In the couple outside of the couple, this strange mother who gives place without engendering can no longer be considered as an origin. She/it eludes all anthropo-theological schemes, all history, all revelation, and all truth. Preoriginary, *before* and outside of all generation, she no longer even has the meaning of a past, of a present that is past." Derrida, "Khôra," 124–25.

92. The title *Feu la cendre* (1982) is translated to *Cinders* in English.

93. Foucault writes, "The obstinate murmur of a language talking *to itself*—without any speaking subject and without an interlocutor, wrapped up in itself, with a lump in its throat, collapsing before it ever reaches any formulation and returning with a fuss to the silence that it never shook off. The charred root of meaning." Foucault, *History*, xxxi–xxxii.

94. See Derrida, "Cogito et histoire," 464.

95. "I write between two resurrections, the one that is given then the one that is promised, compromised to this almost natural monument which becomes in my eyes a sort of calcinated root, the naked spectacle of a photographed wound—the bedsore cauterized by the light of writing, to fire, to blood but to ash too." Bennington and Derrida, *Jacques Derrida*, 54. Hélène Cixous also cites this passage in her celebration of Derrida, *Portrait of Jacques Derrida as a Young Jewish Saint:* "J'écris entre deux résurrections, la donnée puis la promise, compromise à

Derrida, . . . the cinder—leaving nothing, finally, to reveal or unveil—names the absolute non-memory which burns at the origin of language."[96] Finally, in his botanically titled *Dissemination* (1972), Derrida pursues these same origin-related problems using a more textually experimental writing practice. In his introduction, he begins with the problem of paleonymy, the summoning up of old names in new circumstances, writing self-consciously about the form in which his text will present itself.[97] Again, the problem of linguistic indebtedness arises. Why must old names stay? How long must the past be given credit, renewed, recycled, appeased? The problem seems to arise from the conflict mentioned earlier between diachronic and synchronic considerations of language. If language is taken out of time—such an extraction would be a merely virtual exercise—how are we to think of ourselves in relation to these words?[98] Why does old consistently mean better, more authentic, more pure? Here, Derrida's position aligns well with that of Paulhan. Those who aim to legitimate etymology are unwilling or unable to acknowledge the nonrational motivations of their appeal to word roots. The titular concept of Derrida's book is dissemination, which "endlessly opens up a *snag* in writing that can no longer be mended, a spot where neither meaning, however plural, nor *any form of presence* can pin/pen down [*agrapher*] the trace" (26). It is a dis-semination, an undoing or a disruption of the "semination" at work in all texts. In the preface, or "hors-livre," he calls a preface a *seed* that casts a paternal shadow over the entire text.[99] The language of psychoanalysis imposes itself here, with the

---

ce monument presque naturel qui devient à mes yeux une sorte de racine calcinée, le spectacle nu d'une blessure photographiée—l'escarre cautérisée par la lumière de l'écriture, à feu, à sang mais à cendre aussi." Cixous, *Portrait*, sec. 10, no page numbers.

96. Wortham, *Derrida*, 23.

97. "While the question of the 'book' is now going through a period of general upheaval . . . the book form alone can no longer settle—here for example—the case of those writing processes which, in *practically* questioning that form, must also dismantle it. Hence the necessity, today, of working out at every turn, with redoubled effort, the question of the preservation of names: of *paleonymy*. Why should an old name, for a determinate time, be retained? Why should the effects of a new meaning, concept, or object be damped by memory?" Derrida, *Dissemination*, 3.

98. "The presumed origin of a concept or the imagined etymology of a word have often been held up against the process of their transformation, without any regard for the fact that what was being utilized was precisely the vulgar sign most heavily overladen with history and unconscious motivations." Ibid., 182n8.

99. "[The preface] is the word of the father assisting and admiring his work, answering for his son, losing his breath in sustaining, retaining, idealizing, reinternalizing, and mastering his seed. The scene would be acted out, if such were possible, between father and son alone: auto-insemination, homoinsemination, reinsemination." Ibid., 44–45.

archetypes of father and son hovering in a closed sexual loop. The purpose of Derrida's emphasis on this redefinition of dissemination, not as a spreading of seed but as a de-seeding, is to illustrate a method for calling into question all inherited assumptions about the written word. Dissemination, like deconstruction, seeks to uproot old names and old words, to unseat and unseed the father.[100]

His texts, with their noncommittal and evasive techniques of endless deferral, have left many readers throughout the decades unsatisfied, if not utterly indignant. He has been called an ironist, a charlatan, an irresponsible thinker, and countless other names by some, and celebrated as a saint and savior of philosophy by others. One has to wonder why his texts produced so much simultaneous and extreme anxiety and praise. If his ideas were ridiculous, why were they not simply ignored? What is so appealing or repulsive about Derrida's ideas to newer generations of scholars who have arrived long after the first heat of those early debates in the late 1960s had time to cool? Because Derrida was very much a product of the phallogocentric system he critiques,[101] we could borrow an image Bachelard uses in his chapter "Roots" as an appropriate metaphor for Derrida's oeuvre: it is the root that uproots.[102] Whereas his philosophical "fathers" Heidegger and Husserl sought to "reroot" philosophy, Derrida sought to uproot it altogether. The appeal for some, particularly those disenfranchised by phallogocentric thinking, is Derrida's critique

100. "Once again we confront the business of the old name, the onymism in general, of the false identity of the mark, all of which dissemination must disturb at the root." Ibid., 45.

101. Derrida describes a favorite of his targets, phallogocentrism, as an "old and enormous root": "Freud, like those who follow him here, only *describes* the necessity of phallogocentrism, to explain its effects, as obvious as they are massive. Phallogocentrism is neither an accident nor a speculative mistake attributable to this or that theorist. It is an old and enormous root we must take into account." Derrida, *La carte postale*, 509. In *A Thousand Plateaus*, Deleuze and Guattari concede. Of psychoanalysis, they write, "[Psychoanalysis] subjects the unconscious to arborescent structures, hierarchical graphs, recapitulatory memories, central organs, the phallus, the phallus-tree—not only in its theory but also in its practice of calculation and treatment." Deleuze and Guattari, *A Thousand Plateaus*, 17. They are critical of Freudian psychoanalysis throughout *Thousand Plateaus*.

102. Bachelard writes that the first plow was made from a root and that thus this root was used to uproot other roots: "la charrue-racine déracine les racines." Bachelard, *Terre*, 297. In a note, Martin Hägglund writes of his use of the term "radical" in the title of his book *Radical Atheism: Derrida and the Time of Life*: "My use of the term 'radical' . . . does not point to a unitary root or ground but seeks to demonstrate that the root uproots itself and the ground undermines itself. Radical atheism goes to the 'root' of the religious conception of desire in order to show that it is divided against itself." Hägglund, *Radical Atheism*, 207n1.

of a certain kind of conservatism, conservatism in the sense of the will to preserve tradition. As I have shown throughout the preceding chapters, a belief in the redemptive power of roots is nearly always motivated by conservatism, be it genetic, cultural, intellectual, or political. So in the case of this young philosopher who arrives, publishes three books in one year, each of them making bold claims about the grounds on which the institutions that trained him are built, one can see why nearly everyone in the academy suddenly became a conservative in light of his claims and felt compelled to respond to his provocation. Many of those scholars sanctioned by the institution were insulted, while many looking in from the outside were delighted. It was no longer this or that school of philosophy that could be accused of conservatism, but philosophy itself. Given his success in the United States, one could take this as another instance of the younger nation trying to defy and define itself in opposition to Old Europe, where the distant roots of American philosophy were deeply embedded. Derrida, this sower of discord, divided the academy in America and in Europe, causing scholars to ask one another in hushed tones, "Are you a Derridean?," the very adjective incendiary enough to launch hours-long debates involving high personal stakes. After its peak (in the 1970s or 1980s, depending on variable criteria), his legacy has become an uncertain one, which would certainly please him, since undecidability was among his favorite themes. Whether one reads his texts as literature, as philosophy, or as a combination of both, his impact on all fields of the humanities is irrefutable. He succeeded in changing the conversation.

BLANCHOT AND THE ETYMON'S DANGER

> Why does filiation impress us so?
>
> MAURICE BLANCHOT, *The Writing of the Disaster*

Another notable poststructuralist critic of etymology is Maurice Blanchot, who includes himself in a genealogy of etymological skeptics like Paulhan. Cautioning against the abuse of the discipline in a well-known passage from *The Writing of Disaster* (*L'écriture du désastre*, 1980), Blanchot warns of the seduction of filiation and its misapplication to a nonbiologic phenomenon. Blanchot expresses the same reservations as Paulhan regarding the legitimacy of etymology. Essentially dissolving the distinction between academic etymology and popular etymology, he argues that the science-minded etymologist is fooling himself. Despite stringent efforts to prove his legitimacy, his vocation is no more noble than that of a layman who hears uncanny and entertaining resonances between words. What is concerning about the indiscriminate use of etymology is that those who use it seem to have suspended their critical faculty, accepting the validity of their findings without question. Our affective need for a feeling of connectedness to the past trumps all rationality, leading even the staunchest rationalists toward a discipline they know, perhaps secretly, to be purely speculative. In a subsequent passage, he identifies the most troubling aspect of etymology: its implied eschatological conception of history.[103] As we have seen, Levinas argued that Heidegger's texts are rife with pagan mysticism,[104] and Blanchot pursues the connection between root-seeking and mysticism in his identification of the etymological tendency toward a mystical and teleological vision of history. The beginning of words leads us to the end of days. Even the most rational of thinkers are liable to turn toward the esoteric *etymon* but cannot acknowledge that such a turn is illogical—or, perhaps, prelogical. Blanchot writes plainly, "What attracts us to etymology is its unreasonable part more than what it explains: we are interested by the form of enigma that it preserves or doubles as it deciphers" (106). He goes on to argue that writing guarantees itself a position as the trove of secrets (119–20). In addition to its enigmatic promise, etymological research seems to provide a cure for the feeling of existential homelessness. The primordial word cannot be demolished and thus invites all users of a particular idiom to rediscover and dwell in it, to become newly rooted in a remoteness that had been nearly forgotten (86). Burgeoning is a figure for futurity; the allegorical root always contains within it the possibility of a new *generation*. In French, the term for *etymon*, or original word, is the *mot souche*, translated literally as "stump word," but a stump that contains the possibility of life inside it, like the stem cell (*cellule souche*). It is a stump, but not a dead one. Blanchot goes on to illustrate the circular logic of the appeal to etymology, the "circularity from which all languages draw their fecundity," and shows the centrality of botanical tropes in this logic.[105] Blanchot

103. Blanchot, Writing, 97.

104. See "Heidegger, Gagarine et nous," in Levinas, *Difficult Freedom*.

105. The root having been named by analogy with plant growth and with the supposed unity of a germinating principle hidden beneath the earth, we draw from there the idea that the root is the formative germ by which words, in diverse languages, receive power of development, and creative enrichment. Once again, believers and nonbelievers are both wrong and right. A writer who, like Heidegger, goes back to the root of certain words which are said to be fundamental, and receives from this root an impulse to develop variations upon certain thoughts and words, makes the idea that there is in the root a strength that works and that incites to work, "true."

Blanchot, *Writing*, 107

yields momentarily, acknowledging at least one benefit of a certain kind of etymologizing. The minor saving grace of creative etymologies, their encouragement of the thinker to "develop variations upon certain thoughts and words," emphasizes the expressive potential of the discipline, which stands in opposition to its truth-reviving imperative. But this slight "rightness" of the more creative-minded etymologizers ultimately does not redeem the discipline for him. In Blanchot's extended critique of the human as an "etymologizing animal," he seeks to expose this pattern as a human epistemological habit, a curious need to know things that have been confirmed as unknowable. While Michel Leiris has argued that "the everyday meaning and the etymological meaning of a word can teach us nothing about ourselves, since they represent the collective part of language, which was made for all people and not for each of us individually,"[106] Blanchot argues that the collective search for (word) roots tells something quite tangible about humans, namely that by representing the past through a clean and linear textual story (an etymological reconstruction, a written testimony, a history book), the past is made provisionally manageable; however, nothing about such a story precludes a repetition of the same tragedies again and again. In fact, writing used in this way seems almost to precipitate the recurrence of such events. Already the format of Blanchot's book, a series of disjointed prose fragments, challenges the way histories and arguments have typically been constructed as cohesive and logical narratives or reasonings, regardless of what they attempt to communicate. In a move similar to the one seen in Derrida's later texts, which put into practice his hesitations about the written word and its uncontested hegemony, Blanchot refuses to write about collective trauma—in the form of the Holocaust or of other specific or unspecific forms of disaster—using the conventional formats proffered by literature, philosophy, and history. It makes sense that he takes etymology to task, since this discipline relies on a similar myth of cohesion. The tropes and conventions that surround the telling of disaster through written language are unsettled in his text, making clear that Blanchot's indictment of etymological reconstruction is inseparable from his indictment of uncritical reconstructions of the traumatic Event. The abuses of etymology should matter to us, he

Often, etymologists themselves push the botanical metaphor of the word root to its extreme, such as in the title of James Mitchell's 1908 book *Significant Etymologies: Roots, Stems, and Branches of the English Language*.

106. In his *Glossaire,* under the heading "étymologie," Leiris juggles syllables: "homme (mythe, élégie), l'horloge des mots t'immole!" (man [myth, elegy], the clock of words immolates you!). Leiris, *Glossaire,* 87.

argues, because they follow a logic similar to the one that guides the (re)writing of the Event into an innocuous format, a clean and safe representation that makes it easier to accept as a story and as a history, and as cleanly dismissible as a fictional one.

*

This chapter has brought attention to language as the site of root-seeking. Not only is the human body regularly botanified; so are thought—shown in chapter 2 through the examples of Ponge and Valéry's vegetal depictions of thinking—and language as well. Analogies with the plant world are seen in every aspect of what makes the human unique as a species. We may add etymological genealogy to the field of biologic genealogies, genealogies of thought, and transmission of land and cultural heritage as yet another site of interrogation into the relationship between the present and the past. Language is particularly enticing as such a site because, while it is a part of everyday life and the necessary medium through which we communicate, its most distant roots are absolutely untraceable, a blank surface upon which countless narratives may be projected, narratives that guarantee power to some and powerlessness to others. I have shown how, through the very medium of language, various thinkers have attempted to sequester its unknowable aspects and argue for the primacy of this or that culture over all others. This is accomplished most handily through a mysticized chronicle of a word. Words are cast as humans with supernatural births, with living genealogies, and their biographies accentuate the most remote and inscrutable aspects of their lives. The holy bodies of these words are ripe for exhumation, a rebirth after a subterranean interval. To use the Icelandic musician Björk's expression, the reconstruction of etymological origins is an attempt to recuperate a lost "pagan poetry."

The mystery, the enigma, words invested with arcane meaning, the beginning and end of all things: despite Europe's ostensible advancements through the Enlightenment project, the etymological obsession that marks the twentieth century (and the twenty-first as well) is evidence of an irrepressible attraction to Endarkenment.[107] The mechanics of this attraction are complicated. A search for roots seems at first to be an "enlightened" or "enlightening" search for truth, an attempt to find logical arguments to explain or undo certain

107. Although I'm aware that the term "Endarkenment" is a pre-twentieth-century term, I discovered it through the title of Texan musician Ray Wylie Hubbard's 2010 album "A. Enlightenment B. Endarkenment Hint: There Is No C."

patterns in human behavior. By finding what has been forgotten in language, perhaps more sense can be made of the conduct of the species. Thinkers who use etymologies or other forms of root-seeking believe themselves to be involved in a rational scrutiny of the beginnings of culture. However, from the examples I have provided, it is clear that those who engage in such investigations are often predisposed to mysticism and to an understanding of the world in terms of latent, supernatural codes. Root-seeking is a prerational impulse, a conviction that origins are pure and that beginnings are better. The gesture of exhumation (an exposure, from the depths, to the light of truth) implied in etymological research is, in actuality, a form of occultation.

Jean Starobinski articulated what is perhaps the best formulation for dealing with the past without surrendering to mystification. Instead of a binary choice between an ignorance of the past and an irrational indebtedness to it, he proposes this solution: "There is no reason . . . why our interest in the cultural past should diminish if, instead of representing a part of ourselves, this past consisted in things other men have accomplished within a conceptual framework which is not and will never be ours, using a language in which we recognize nothing of ourselves."[108] Starobinski cuts the imagined umbilical cord between contemporary and ancestral culture. Because it is virtually impossible to locate a true cultural continuity between ourselves and the most remote of our ancestors, the imaginative and superstitious mind would rather fill in the gaps of history with fantasy in order to remain a part of it. The fabulator would do better to dispense altogether with this tendency, suggests Starobinski, admitting that we have no meaningful connection to those who came before us other than a symbolic one. People are possessed by and possessive of their ancestors, but this is a treatable condition, in his view. By severing the imaginary bind and opening up the mind toward multiple histories rather than one's own singularly important (and mostly fabulated) History, the hazardous aspects of ancestor worship, such as the sacralization of the banal and the celebration of the past to the detriment of the present, could be neutralized. As I showed in chapter 4, Vilém Flusser pursues synchronically what Starobinski has undertaken here diachronically; while Starobinski confronts the dangerous sacralizing gesture of ancestor fetish, Flusser confronts the dangerous and timeless sacralization of the homeland.

The ways in which etymology was instrumentalized throughout the twentieth century—particularly in France and Germany, the epicenters of consequential shifts in the root metaphor—show in reduced scale the broader

108. Starobinski, "Inside," 335.

tension that prevailed in various discourses during the period, namely, the tension between an etiological worldview, a teleological one, and other narratives of human agency. An etiological viewpoint thinks from the beginning, looking toward the origin for answers. A teleological one thinks from the end backward, looking toward finalities and "destinies" for answers about human trajectory and purpose. The two are not mutually exclusive. One can easily create a narrative that claims that the beginnings and the futures of the human are part of a well-designed plan. Dissenting voices include those that dismiss an etiological and teleological view of culture altogether, and those that tweak the idea of indebtedness either to the past or to the future, claiming only a partial responsibility of the human to his forebears or to his progeny. His present is shaped in part by past and future, but he can act in accord or discord with these pulls.

In the following chapter, I will show how the *vocabulary* used to describe ways of knowing changes throughout the century while, in reality, the difference between so-called modern and postmodern epistemologies remains the same. Again, this shift relies on a botanical vocabulary, a move from a root-based epistemology to a rhizomatic one, using the metaphor offered by Gilles Deleuze and Félix Guattari. Their metaphor has had lasting appeal, particularly in the fields of philosophy, literature, history, and the social sciences. Confusion still remains, however, over whether such a seemingly "groundbreaking" concept has actually destabilized old disciplines and methods or whether the category of the rhizomatic is an extension of "rooted" epistemologies.

CHAPTER 7

From Rhizome to Vegetal Democracy

If it is true—as Ihab Hassan, Emily Apter, and others have asserted—that the twentieth century represents a shift from a root-based epistemology to a rhizomatic one, one wonders whether the current call for vegetal democracy is a continuation of this "rhizomization," or whether it is something else entirely. I don't believe the new attention to plant life, as articulated in the work of thinkers like Matthew Hall, Eduardo Kohn, and Michael Marder, could have been possible without the strange path of the root metaphor I analyze in this book. As I will show, Gilles Deleuze and Félix Guattari retained in their writing on the rhizome some of the mystical features of writers like Claudel, Heidegger, and Weil, which were passed on in turn to the new proponents of vegetal democracy. Furthermore, there is a shift in narrative, from a rejection of rootedness to what could be called "rootedness in rootlessness," the attempt to find comfort in a world that has disposed of context. The overlap of politics and ecology, of human and plant, have led to the attempt to grant certain democratic rights to plants, an animistic gesture that invites paganism to infiltrate politics. Rhizome-centered thinking, which breaks the root and the power structures it represents into an unpredictable, undecidable abstraction, created the conditions under which the human loses its status as the universal sovereign over all other forms of life. In this final chapter, I tie together various strands that have run through this study, including the place of nationalism in our postnational moment, the place of metaphysics in our postmetaphysical moment, and the place of humanism in our posthuman moment.

This chapter identifies Deleuze and Guattari's work as the most important connecting link between the early-twentieth-century paganistic uses of the plant-human nexus and the contemporary postnational, postmetaphysical, and posthuman tendencies in philosophy. I argue that their thought works as a kind of checkpoint through which several lines of rooted thinking passed and transformed. For example, on the identarian front, the root thinking of early-twentieth-century nationalists such as Barrès refracts through Deleuze and Guattari's rhizomatic and nomadic metaphors and finds a place in the work of thinkers like Edouard Glissant and Antonio Benítez-Rojo, who identified the prescient Caribbean as the true site of a certain kind of laudable postmodern chaos.[1] Their application of the rhizome metaphor, among other image concepts from *Anti-Oedipus* and *A Thousand Plateaus*, leads them to place a great amount of emphasis on the plantation as the primary structuring element of Caribbean culture. The plantation is where the human and the plant meet and where economic and social futures are shaped. Nature and culture intersect here. Deleuze and Guattari provided a visual language through which these various intersections could be sometimes fruitfully analyzed. Regarding the mystical strand of plant-centered thought, I show that a heavily mystical element is maintained, from those theories of rootedness formulated by Claudel, Heidegger, and Weil, then through the rhizome and nomad figures of Deleuze and Guattari, and, finally, to the mystically infused posthuman ontologies of thinkers like Matthew Hall, Eduardo Kohn, and Michael Marder, to mention only those who propose the plant world specifically as deserving of new focus. In other words, I attempt to make a case that a certain line of magical thought has led from Claudel and Weil's Christian and Heidegger's paganistic metaphysics through Deleuze and Guattari to the current proponents of vegetal democracy and plant thinking.

In the final thoughts this book has to offer, I show how a hyperliteralization of the root metaphor has resulted from currents in poststructuralist and postmetaphysical thought and in new research in the fields of botany and anthropology. This literalization has led to the conclusion that the human-as-plant metaphor was never a metaphor. Humans and plants have always been in direct communication via their shared cellular consciousness, which is so intimately and reciprocally attuned that the boundaries between plant and human are dissolved. This theory explains our attraction to plants, our endless botanification of people and anthropomorphization of plants, and our

1. For more on Deleuze's influence in the Caribbean, see Lorna Burns, *Contemporary Caribbean Writing and Deleuze: Literature between Postcolonialism and Post-Continental Philosophy* (London: Continuum, 2012); and Lorna Burns and Birgit Kaiser, eds., *Postcolonial Literatures and Deleuze: Colonial Pasts, Differential Futures* (London: Palgrave Macmillan, 2012).

centuries-long insistence on the rootedness of humans. Matthew Hall's *Plants as Persons: A Philosophical Botany* (2011), Eduardo Kohn's *How Forests Think: Toward an Anthropology beyond the Human* (2013), and Michael Marder's *Plant-Thinking: A Philosophy of Vegetal Life* (2013) are new poststructurally influenced reworkings of Peter Tompkins and Christopher Bird's *The Secret Life of Plants* (1973), which argued that plants merit the same consideration we give willingly to other humans and to some animals. We should behave more like plants, they argue, taking only what we need, slowing down our pace, and operating on the ethical principle of trans-species dignity.

THE CRYPTIC RHIZOME OF DELEUZE AND GUATTARI

After reading *A Thousand Plateaus* several times and ruminating long and hard, I am still not certain I know what a rhizome is. If poetry and philosophy are the same thing, I can accept this; the rhizome certainly adds a dose of poetic appeal for some theory seekers. But if they aren't the same thing, I have to wonder why the botanical metaphor that I've followed throughout the book with assurance has dissolved in the hands of Deleuze and Guattari. I also must wonder why recent books that argue in favor of vegetal democracy or plant-human empathy, books directly influenced by poststructuralism and rhizome-centered thinking, are as problematic as the nationalist claims of Barrès and the mystically infused autochthony of Heidegger. The botanical thought that connects all of these disparate texts and theories always brings strangeness with it.

There were moments when I believed I understood the rhizome. I thought it might be a metaphor for describing those phenomena that refuse hierarchy, linearity, and institutionality. I imagined it as an improvisational form that communicates the idealized unmanageability of transmitted culture. It seemed to be a living alternative, something that shared certain features with roots or rootlets, such as the fact that these are all botanical forms that grow underground and are alive and mutable. The rhizome concept seemed to want to provide a language for talking about those forms of thinking and living that do not fit into clear categories or that have not been sanctioned by the state or other centers of power. But as I did more research on how the rhizome has been used by others, especially social scientists, I realized that the concept can be made to mean almost whatever one wants it to mean.[2] There is certainly

2. Here is a sample of titles to give a sense of what I mean: "Becoming Rhizomatic Parents: Deleuze, Guattari and Disabled Babies" (Goodley, 2007); "Disciplinary Power, the Oligopticon and Rhizomatic Surveillance in Elite Sports Academies" (Manley et al., 2012); "The Rhizomatic

something appealing about the alternativeness and malleability of this concept and Deleuze and Guattari's other nearly synonymous concepts expressed alternately in their corpus through the language of ethnography/anthropology (the nomad), biology (the organless body), and martial mechanics (the war machine), to name just a few. I am certainly sympathetic to a project that would multiply concepts through which the world may be thought, especially when the ultimate objective is to move toward a less West-centric, white-centric, male-centric consideration of this world. However, one must wonder whether the rhizome is so radically different from the root in the end. Ultimately, its application to various discourses does not differ greatly from the application of the root metaphor to these same discourses.

Among recent attempts to turn Europe's roots into rhizomes, or at least to locate them in unexpected places, is the recent push to locate the roots of poststructuralism in North Africa.[3] Given that many of the key players responsible for the development of poststructuralist thought were either from Africa (Althusser, Derrida, Cixous) or had been impacted heavily by their time in Africa (Leiris, Foucault, Bourdieu, Lyotard), the notion of "French theory" is destabilized and transformed into "African Theory" or even "Algerian Theory," "Maghrebine Theory," "French-African Theory," "Jewish-African Theory,"

Relations of A/r/tography" (Irwin et al., 2006); "Organizing and Regulating as Rhizomatic Lines: Bank Fraud and Auditing" (Bougen and Young, 2000); "A Rhizomatics of Hearing: Becoming Deaf in the Workplace and Other Affective Spaces of Hearing" (Crowley, 2010); "Rhizomatic Explorations in Curriculum" (Smitka, 2012); "Rhizomatic Thought in Nursing: An Alternative Path for the Development of the Discipline" (Holmes and Gastaldo, 2004); "COMMUNEcation: A Rhizomatic Tale of Participatory Technology, Postcoloniality and Professorial Community" (Broadfoot et al., 2010); "From Cluster F&%K to WTF: A Rhizomatic Reading of 4Cs Area Clusters in Technical and Professional Writing" (Smith, 2011); "Reuben's Fall: A Rhizomatic Analysis of Moments of Disobedience in Kindergarten" (Leafgren, 2007); and "Rethinking 'Foster Child' and the Culture of Care: A Rhizomatic Inquiry into the Multiple Becomings of Foster Care Alumni" (Corcoran, 2012).

3. See, for example, Robert Young, *White Mythologies: Writing History and the West* (London: Routledge, 1990); and *Postcolonialism: An Historical Introduction (*Oxford: Blackwell, 2001); Pal Ahluwalia, *Politics and Post-colonial Theory: African Inflections (*London: Routledge, 2001); Christopher Wise, *Derrida, Africa, and the Middle East (*New York: Palgrave Macmillan, 2009); and Ahluwalia, *Out of Africa*. Ahluwalia writes, "It is my contention that, in order to understand the project of French post-structuralism, it is imperative both to contextualise the African colonial experience and to highlight the Algerian locatedness, identity and heritage of its leading proponents. It is precisely the failure to confront or explicitly acknowledge the colonial experience that problematises the conflation of postcolonialism and post-structuralism." Ahluwalia, *Out of Africa,* 3.

and so forth, depending on which theorists are invoked as founders of poststructuralism. What happens to the notion of late Continental philosophy if the origin of ostensibly Continental European ideas in the twentieth century is located not on the European continent but on the African one? This attempt to locate the origin of supposedly European ideas outside Europe is another method for undermining narratives of origin. There seems to be a collective hope that by relocating origin, self-perpetuating Eurocentrism can be—at least provisionally—disrupted. By showing that ideas are never roots but always rhizomes, or that they have no localizable beginning, or that their beginnings are never neatly national, the European is forced to concede a loss of intellectual territory. However, even if the root of contemporary Western thinking is not Western, this does not change the fact of the compulsive search for roots. The attempt to Africanize and nomadize European thought ultimately does not undo its arborescent nature; if anything, it reinforces it. Roots are a given.

While Derrida tried to set up the conditions for Western thought to uproot itself, Deleuze and Guattari provided alternative metaphors through which all *idées reçues* could be reworked. This move is similar in structure to one made by the Oulipo when they asked why we should be satisfied with the old literary constraints—like the strictures of the sonnet form, for example—and why we haven't made more efforts to add new ones rather than repeat the somewhat arbitrary constraints drafted by our forebears. Similarly, Deleuze and Guattari wonder in their pages why we have taken up the old arborescent models with zero reluctance. Why this obsequious indebtedness to the past? Why this indolent compliance with forms that may have expired? The world has the potential to be thought through a different set of images, they posit. Despite their claims that the concepts they propose (rhizome, war machine, nomad, organless body, etc.) are not metaphors, readers cannot but take them as such. The attempt to accept the authors' claim and read these figures literally unsettles the mind with "great doubt," much like the exercise of Zen kōan contemplation. The figures somehow undo themselves when imagined through the lens of the literal. Assuming that they actually are metaphors, what is most captivating about them is their imagistic quality. Deleuze and Guattari attempt, through their literary essayism, to rethink the world using a new set of written pictures. Their grievances against Western arborescence are addressed imagistically rather than textually, despite the written format of their books. In a manner compatible with Derrida's frustration over the autocracy of writing, Deleuze and Guattari use text only as a means to build up a cache of self-multiplying images.

At times, their metaphors seem to be a mere multiplication of synonymous images and a series of non sequiturs. To get out of this closed circle of interpretation, I try to look more panoramically at how they use metaphors to make their arguments. Deleuze and Guattari's work offers an excellent occasion for an analysis of philosophers' appropriations of metaphors. Throughout this book, I have shown how the malleability of the root metaphor allows it to mean one thing, its opposite, and much in between. For some, it is a figure of restrictive immobility, for others a figure of necessary stability. It may represent continuity between past, present, and future, or a stubborn attachment to the past and a refusal of all that comes after. If a metaphor can be massaged into meaning almost anything, what use is it to anyone? What use is a metaphor to a philosopher?

Stephen C. Pepper has suggested, "Every philosophical theory is a far-flung metaphor."[4] At times, the rubbery nature of metaphorical language can be philosophy's strength, particularly since the field focuses so frequently on abstractions. In large part, metaphor allows for an escape from the kind of responsibility that apodictic, metaphor-free language seems to require. When one expresses an idea with an apodictic sentence, void of metaphor—which is the ideal of many philosophers, albeit nearly impossible to achieve—readers are left with little interpretive opportunity. This form of philosophical expression seeks to reduce occasions for misunderstanding and to relieve the philosopher in advance from needing to reiterate, justify, or expound. Perhaps there is readerly philosophy and there is writerly philosophy, to use Barthes's distinction. The books of Deleuze and Guattari are clearly writerly texts. Their reliance on metaphor seems to be an open invitation to improvisation of meaning on the part of the reader.

For this reason, I claim that the greatest contribution of their collaborative work belongs to the field of poetics. Read through the lens of literature, *Anti-Oedipus*, *A Thousand Plateaus*, and *What Is Philosophy?* are image-inflected prose improvisations. These books are imagery generators with a goal of self-proliferation. They illustrate the possibilities of an aesthetically conscious kind of writing that is also philosophically, politically, and historically informed, clustering bits of knowledge drawn from these fields around imagistic figures. They do not proceed by argumentation; they proceed by image generation, which continues to operate in the reader even when the pages have all been read. I believe it was a conscious choice on the part of Deleuze and Guattari

4. Pepper, "Philosophy and Metaphor," 130.

to make a book that sidesteps the strictures of theory or philosophy, relying instead on the literary and visual imagination of readers to do most of the books' work. Like Robert Musil and Hermann Broch's "novels that think,"[5] which kept expanding various lines of thought, all within the frame of fiction, the collaborative books of Deleuze and Guattari keep going, fueled by an ever-growing stock of "images that think." A reliance on images was obviously not new to philosophy or theory, but the method of their unfurling in these books, imitative in many respects of Derrida's image work, tries to mimic the same style of movement as the figures they describe in words (the rhizome and the nomad, for example). These image-centric books are nonetheless thinking texts, literary essays that abound with images that generate other images outside the pages, for example in Edouard Glissant's appropriation of the rhizome figure and his transformation of it into an archipelagic, relational figure. The collaborative books of Deleuze and Guattari are image factories, or, perhaps a better term, image plants. By spinning out and echoing an endless series of synonymous images that "mean" in the same way, they write a kind of image-reliant thesaurus or picture book of synonyms. In this sense, Deleuze and Guattari engage in a practice similar to that of Paul Claudel, who, as I showed in chapter 3, delivers his proselytizing message by inundating his reader with image after image of the symbols of Christian love, combined in every possible configuration. As I will show, they also share a penchant for mysticism. The surrogate universe these three writers create thinks our own world through a different optical apparatus. By replacing argument with a flurry of pictures, they imagine they'll change minds, since the human is such a visual creature. We no longer choose and commit to one picture; we accumulate and shuffle many.

The appeal of the rhizome concept has continued in many disciplines because, in its preference for imagistic poetics over rhetorical argument, it has the feel of a true expressive alternative. However, replacing the old botanical metaphor with a new one does not radicalize thought in the way Deleuze and Guattari imagined. In their theorization of the rhizome, they sought to unsettle philosophy from its ground, using gestures similar to those of Derrida.[6] But

5. Kundera, *Le rideau*, 87.

6. Brigitte Weltman-Aron compares the Derridean khôra with Deleuze and Guattari's rhizome: "Both khôra and rhizome name places refractory to the proper, and both give place to another thinking of the time/space of the political. Both exceed all cosmologies, understood as discourses of origin and structure, and both are deployed in relation to writing." Weltman-Aron, "Rhizome and Khôra," 50.

the root-to-rhizome shift is overly touted as a salvific move toward freedom from the past. In the end, the rhizome is not radically different from the root. It is still an embedded form, its pace of growth is the same, and it represents yet another instance of thinking botanically about thought. Their plant-picture puts them in a long genealogy of Western thinkers who conceive of thought in botanical terms.[7]

The rhizome thrives in unpredictable fits and starts below the earth's surface. The nomad skims smoothly if unforeseeably across the earth's surface. In both cases, the earth itself is the necessary precondition for their survival. Everything in our world grows from the ground up. Deleuze and Guattari's coauthored texts are profoundly earthen; that is, they rely constantly on metaphors of ground, earth, geography, geology, and terrestrial unity. For example, from their geocentric preoccupations in *Anti-Oedipus* and *A Thousand Plateaus* to their engagements with Nietzschean geophilosophy in their book *What Is Philosophy?* (1991), they are no less ground-focused than the phenomenologists. In an attempt at an antichronology, each "plateau" of *A Thousand Plateaus* is marked with a year and a title, shuffled out of order. The earliest date they choose, 10,000 BC, is entitled "The Geology of Morals." By replacing Nietzsche's "genealogy" with "geology," Deleuze and Guattari substitute for a vocabulary of human genetic transmission a vocabulary of the terrestrial, perhaps intentionally recalling—through the association of the earlier human and earthly clay—early Greek accounts of the origins of humanity or the Adamic beginning and the subsequent and quick moral fall. Origins are the centerpiece here. In their claim that "Thinking takes place in the relationship of territory and the earth,"[8] they underscore the fact that our ground, the necessary precondition for all terrestrial life and thought, is never out of mind. Their rhizome is textually embedded in the ground, regardless of their attempts to render this ground seismically unstable. And, while the nomad is not fixed in the dirt, he requires its surface as his platform of potential destinations.

The rhizome has proved to be their most lasting contribution inside and outside philosophy. First in 1976 and then again in 1980, Deleuze and Guattari proposed the rhizome as an alternate metaphor to the root for a nonbinary, nonhierarchical epistemology, and since the publication of their pivotal

7. Michael Marder's *The Philosopher's Plant: An Intellectual Herbarium* catalogs various thinkers' appropriations of plants as sites of philosophical contemplation, from Plato's plane tree to Irigaray's water lily. In the introduction, he writes that his book "lifts the curtain on the significance of plants to the making (and growth) of thought." Marder, *Philosopher's Plant*, xiv.

8. Deleuze and Guattari, *What Is Philosophy?*, 85.

text, this metaphor has been appropriated by theorists to describe the shift from modernism to postmodernism.[9] Throughout *A Thousand Plateaus* and their other coauthored works, whose method is described by Alexandra Kogl as "psychedelic,"[10] Deleuze and Guattari sought to fractalize the ostensibly cohesive and arborescent form of Western thought. Because fractals are by nature centrifugal, their particular method of image-making quickly spread across disciplines and national borders. The rhizome concept, for example, has greatly influenced writers across the globe, such as Edouard Glissant and Antonio Benítez Rojo in the Caribbean, who took up the call in their own writing for a celebration of multiplicity, of dissolved categories, and of the horizontalization of all aspects of life. Deleuze and Guattari's metaphor has influenced urban planning and architecture, film studies and filmmaking, music, design, literary studies, philosophy, history, political science, gender studies, and most branches of the social sciences. It is difficult to name a single concept that has had more influence on recent research in all of these fields.

With its emphasis on authorial faltering, limitations, and deficiencies, *A Thousand Plateaus* is a performance of self-doubt.[11] It is also an evasive text, *un texte qui s'esquive*; as Christopher Miller has noted, "Catch the authors in a contradiction, and they will merely dissolve the terms that created it, then take off on another 'line of flight.'"[12] At times, the book's introduction reads like a manifesto to the self, a posted reminder of the self-constraints to be respected.[13] Tipping over into full manifesto mode in the end of the introduction, Deleuze and Guattari write: "Make rhizomes, not roots, never plant! Don't sow, grow offshoots! Don't be one or multiple, be multiplicities! Run lines, never plot a point!" (24). The directives continue. Alternately a manifesto, a diary of doubt, a theory-laden academic document, a fiction, a history book, an essay, and many other genres, the text ultimately refuses to decide what it wants to be,

9. Deleuze and Guattari published the introductory chapter of *A Thousand Plateaus* first on its own, as *Rhizome: Introduction* (Paris: Editions de Minuit, 1976).

10. Kogl, *Strange Places*, 57.

11. Deleuze and Guattari cannot even convince themselves of a sure readership. They write of their text intended for the *peuple*, the masses, who will likely never read it: "RHIZOMATICS = POP ANALYSIS, even if the people have other things to do besides read it, even if the blocks of academic culture or pseudoscientificity in it are still too painful or ponderous." Deleuze and Guattari, *A Thousand Plateaus*, 24.

12. Miller, "Postidentarian Predicament," 9.

13. In the infinitive they use in French, which functions here as an imperative, they write, "Never send down roots, or plant them, however difficult it may be to avoid reverting to old procedures." Deleuze and Guattari, *A Thousand Plateaus*, 23.

which is in itself one of the central motives of the project: to refuse the cleanliness of categories. This poetics of accumulation necessarily results in an unwieldy book, a sedimentary bed where concepts and their opposites amass. This rhetorical rockiness comes through in each and every plateau of the book, leaving many readers at a loss when faced with its lithic opacity.

Probably because the introductory chapter is entitled "Introduction: Rhizome," readers of *A Thousand Plateaus* often walk away from this book taking only the handsomely packageable concept of the rhizome, which can be applied to a variety of fields. If we look closely at this chapter and at the text as a whole, we find countless other metaphors for books of various types.[14] The rhizomatic book is just one among them. Ultimately, *A Thousand Plateaus* relies much less on botanical metaphors than on geological, geographic, and topological ones. This fact is announced already in the titular "plateaus"—a term they acknowledge was borrowed from Gregory Bateson, cyberneticist *extraordinaire*—as well as the constant references to stratification, sedimentation, territory, landscape, cartography, the subterranean, and the earth.[15] They write in the treatise on Nomadology segment that "the earth is before all else the matter upon which the dynamic of lineages is inscribed, and the number and the means of inscription: the lineages write upon the earth and with the number, constituting a kind of 'geodesy'" (388), that is, the mathematical measurement of vast spaces and the numerical taking into account of their shapes. As I have shown in chapter 3, transcendence relies on an immanent ground that must be overcome. In *What Is Philosophy?*, published more than ten years after *A Thousand Plateaus*, Deleuze and Guattari dedicated an entire chapter to the topic of geophilosophy, claiming that "thinking takes place in the relationship of territory and the earth."[16] Following Nietzsche, who

14. These include a machine, a tracing, a container, the world, a field of representation, a flow, and the State, among others.

15. Deleuze and Guattari open their conclusion with this sentence: "The strata are phenomena of thickening on the Body of the earth, simultaneously molecular and molar: accumulations, coagulations, sedimentations, foldings." *A Thousand Plateaus*, 502.

16. Deleuze and Guattari, *What Is Philosophy?*, 85. For them, "The earth is not one element among others but rather brings together all the elements within a single embrace while using one or another of them to deterritorialize territory" (85). In an extended reading of Deleuze and Guattari's geophilosophy, Gregory Flaxman writes, "Etymologically speaking, we propose to lay down a course that leads from a *gene*- qua genealogy to *geo*- qua geophilosophy—from familial and human lineages to the inhuman milieu of the earth. Insofar as genealogy describes our 'all too human' descent, in what sense could it possibly anticipate the movement outside humanity, to 'a world without others' (*un monde sans autrui*)—to the earth? How does genealogy, the

tried to locate the philosophical specificity of French, German, and English thought in *On the Genealogy of Morals*, they puzzle through the role of autochthony in determining the orientation of philosophy. Regarding the Greeks, the distant philosophical forebears who still control European thought even posthumously, they come to the conclusion that "philosophy appears in Greece as a result of contingency rather than necessity, as a result of an ambiance or milieu rather than an origin, of a becoming rather than a history, of a geography rather than a historiography, of a grace rather than a nature."[17] Their sustained interest in "becoming" leads them to a conclusion, drawn from Schelling's and Hölderlin's depiction of the Greeks, that, invariably, "what was Autochthonous becomes strange; what was strange becomes Autochthonous" (102). These categories barely hold up under this constant sway. In contrast to Weil and Heidegger, with their Graecophilia, Deleuze and Guattari do not exceptionalize the emergence of Greek thought and do not tout it as the origin of all origins. They attempt to undermine the temporal aspects of a philosophical filiation by transforming them into spatial aspects. The ground they had surveyed already in *Anti-Oedipus* and *A Thousand Plateaus* loses much of its already precarious solidity in *What Is Philosophy?* Deleuze and Guattari are by no means the only ones to think the world through a geological frame. The human/*humus* nexus plays out in the same period in other venues, such as in Kenneth White's 1989 founding in France of the International Institute of Geopoetics (*Institut international de géopoétique*) and with Jed Rasula's elaboration in 2002 of a necropoetics based on the metaphor of compost, as the processing of dead material (and thought) into new material (and thought).[18]

study of ancestry, presage a geophilosophical practice in which 'paternity does not exist' and the subject dissolves in the slow passage of geological time?" He offers a new definition of genealogy as "a critique of origins that proceeds by foreclosing origins." Flaxman, *Gilles Deleuze*, 73.

17. Deleuze and Guattari, *What Is Philosophy?*, 96–97.

18. In the founding document of the institute, Kenneth White calls for the coalescence of "a major movement concerned with the very foundations of man's existence on earth" ("Il s'agit d'un mouvement majeur qui concerne les fondements mêmes de l'existence de l'homme sur la terre"). L'institut du géopoétique, www.geopoetique.net/archipel_fr/institut/texte_inaugural/index.html, accessed March 20, 2014. White's call for a new movement is marked by a language of fertility, foundation, and return. He conceives it as a borderless community with an interdisciplinary imperative, combining "poetry, thought, and science" and various disciplines like geography, biology, and ecology. Jed Rasula calls necropoetics "a pledge enacted between the dead and the living." Rasula, *This Compost*, 65. Throughout the book, whose full title is *This Compost: Ecological Imperatives in American Poetry,* he uses the figure of composted material—including the bodies of the dead and other matter that was once living, which then transmutes

All of these initiatives take the human as necessarily lodged in the earthen loam of this domicile called Planet Earth.

The rhizome, itself an unstable form, thrives best underground. In contrast to the earth's stony surface, on which histories can be carved and measured, the rhizome's vegetal matter does not last long enough tell a tale and its surface is not dense enough to be inscribed with any lasting message. Any inscription on the rhizome would decompose along with it; it cannot tell History, only Nomadology. Published between the first of Derrida's *trace*-concerned texts and Gérard Genette's 1982 *Palimpsestes: La littérature au second degré* (*Palimpsests: Literature in the Second Degree*), *A Thousand Plateaus* is in sync with the intuitions of Deleuze and Guattari's intellectual cohort. In unison, they wonder how generations preemptively mark those generations to come and note the pivotal role of stone in this process. Vegetal material becomes a surface for inscription only once it has been processed by the human into paper; even then, its durability is inferior to that of stone. The scroll rolls itself up, its message inside, with the same elastic, inward folding of a living plant. In their *Anti-Oedipus*, a book that includes what Emily Apter calls the "presciently postnational vision of deterritorialized subjects,"[19] this terrestrial emphasis was prefigured in the many passages that underscore the role of Earth as the ground from which everything else originates.[20] The rhizome metaphor has been singled out and plucked from the soil of the edaphic text in which it was embedded. But this is to be expected: the authors invited such a singling-out by refusing the totalizing format of a book that presents and sustains itself procedurally. Concepts cannot but be extracted casually and in discrete units from their book, like many a Nietzschean aphorism.

Despite their claims that the rhizome is radically different from the root, the similarities between the two botanical forms far outweigh their dissimilarities. Put in another way, the rhizome allows one to dispose of the root without disposing of rootedness. The rhizome had just as much of an underground

again into living matter—as analogous to the poet's processing of figuratively dead material into the (re)vitalized word.

19. Apter, *Continental Drift*, 227.

20. For example, in a passage on the relationship between labor, production, and the earth, they write, "While the ground can be the productive element and the result of appropriation, the Earth is the great ungendered stasis, the element superior to production that conditions the common appropriation and utilization of the ground. It is the surface on which the whole process of production is inscribed, on which the forces and means of labor are recorded, and the agents and the products distributed." Deleuze and Guattari, *Anti-Oedipus*, 140–41.

beginning as the root; the root is as much "in process" as the rhizome.[21] In fact, in comparing a real root and a real rhizome, the characteristics assigned to them by Deleuze and Guattari seem rather reversed. The root does not respect the neat one-becomes-two bifurcating order they describe, nor do many trees; often, the root is instead a tangled and uneven chaos of fibrils, hairs, and worm-like articulations. Even with its curious contours, the rhizome—in the form of ginger, for example—can be still relatively smooth with a clear-cut shape and a solid consistency in contrast to the enmeshed muddle of the root. Perhaps they would argue that I am taking these forms too literally, but if they really are not metaphors, how else is one to take them? This is not the only contradiction in the book. Despite their attempts to challenge the conventions of bookmaking by suggesting that the various plateaus are readable in any order, they are still explicit that the introduction must be read first and the conclusion last.[22] We could perhaps say that *A Thousand Plateaus* resembles the *livre-radicelle* even more than the *livre-rhizome*, putting them alongside modernists like James Joyce and Nietzsche who, they claim, wrote rootlet-books rather than rhizome-books. In their depiction, rootlet-books function on the principle of cyclical chaos; this applies as well to Deleuze and Guattari's collaborative work, in which the same motifs recur in sporadic fashion.[23]

The subterranean aspects of *A Thousand Plateaus* certainly include the many geological metaphors listed above, but the underground is also exploited in its figurative sense as that which is hidden and enigmatic.[24] Mark Bonta argues that Deleuze "was closely attuned to the subterranean mystical currents

21. Edouard Glissant makes this argument, writing, "The notion of the rhizome would thus maintain the fact of rootedness, but rejects the idea of a totalitarian root." Glissant, *Poétique*, 23.

22. The first section is titled "Introduction," and in the *avant-propos*, they write, "Dans une certaine mesure, ces plateaux peuvent être lus indépendamment les uns des autres, sauf la conclusion qui ne devrait être lue qu'à la fin." Deleuze and Guattari, *A Thousand Plateaus*, 8.

23. "Joyce's words, accurately described as having 'multiple roots,' shatter the linear unity of the word, even of language, only to posit a cyclic unity of the sentence, text, or knowledge. Nietzsche's aphorisms shatter the linear unity of knowledge, only to invoke the cyclic unity of the eternal return, present as the nonknown in thought. . . . The world has become chaos, but the book remains the image of the world: radicle-chaosmos rather than root-cosmos." Ibid., 6.

24. This is particularly clear, for example, in their passages on "animist eyes" (ibid., 211), the animal spirit—"a jaguar-spirit, bird-spirit, ocelot-spirit, toucan-spirit" (176)—and at the end of the plateau entitled "10,000 B.C.: The Geology of Morals," in which they summon up H. P. Lovecraft's "dark, cosmic rhythm which underlies all mystical gate-openings" (74) as an entryway into the rhizophere.

that pervade Western religiosity."[25] Putting him in a genealogy with "a long line of Western intellectuals who have engaged Western esotericism and 'secret knowledge'" (62–63), Bonta approximates Deleuze with Jakob Boehme (1575–1624), the German theologian and Christian mystic who published texts such as *Mysterium pansophicum* (1620) and *Mysterium magnum* (1623), focusing on the origins of the earth and other terrestrial and celestial mysteries.[26] Derrida, too, has been accused, by many of his own brand, of mystification and obscurantism, each of his texts being the unfulfilled promise of a recovery of buried meaning.[27] When Derrida writes, "There has been and always will be philosophical mystification, speculation on the end and the ends of philosophy,"[28] he thinks from the end backward, implying that mystification occurs at both the beginning and the end of thought. As a philosopher, he belongs to the paradoxical discipline that seems to seek truth but unwittingly obfuscates it through an endless series of speculative gestures groping for unknowable beginnings and endings. As I have shown in chapter 1, the subterranean aspect of the root invites a mystical metaphorization. In the debates about the vegetative soul and the inscrutable connection between the human and the plant, poststructuralism and then postmetaphysical philosophy have taken up where mysticism left off. In terms of its botanical content, Deleuze and Guattari's rhizome image provides an important link between the mystical preoccupations about the uprooted human—as found in Claudel's, Heidegger's, and Weil's thought—and the posthuman ontologies that proliferate in philosophy today. Claudel and Weil privilege Christian mysticism, while Heidegger privileges a more paganistic mysticism, although it is easy to make the claim that

25. Bonta, "Rhizome," 62.

26. It is important to note that Boehme also influenced the anthroposophist Rudolf Steiner, Martin Heidegger, C. G. Jung, Martin Buber, Simone Weil, Emmanuel Levinas, and Jacques Derrida, whose names come up regularly in relation to the problem of rootedness.

27. John Searle claims,

> With Derrida you can hardly misread him, because he's so obscure. Every time you say, "He says so and so," he always says, "You misunderstand me." But if you try to figure out the correct interpretation then that's not so easy. I once said this to Michel Foucault who was more hostile to Derrida than even I am, and Foucault said that Derrida practices the method of *obscurantisme terroriste* (terrorism of obscurantism). . . . And I said to Michel, "What the hell do you mean by that?" And he said, "He writes so obscurely you can't tell what he's saying, that's the obscurantism part, and then when you criticize him, he can always say, 'You didn't understand me, you're an idiot.' That's the terrorism part."
>
> Feser and Postrel, "Interview with Searle," 45

28. Derrida, "Of an Apocalyptic Tone," 32.

Christianity never managed to rid itself of the residue of paganism. One can readily identify the paganistic elements that Christianity absorbed into its own metaphysics. The mystery among the trees cannot be uprooted, even in late-twentieth-century philosophy.

Regarding the arguments presented in *A Thousand Plateaus*, many critics have transformed Deleuze and Guattari's trichotomy *livre-racine–livre-radicelle–livre-rhizome* into a dichotomy, focusing primarily on the first and third terms but disregarding the second. This tendency to disregard a Third Term—as a general category, as a symbol for that which throws dualistic thinking off-kilter—illustrates the perennial disdain toward a "complicating element," even in this text that operates according to the principle of proliferation beyond binaries. When Deleuze and Guattari published this antibinary text in 1980, our digital lives were just beginning to get under way. This was the year, for example, when IBM hired Bill Gates and when Apple made its Initial Public Offering in the stock market. Digital life and digital death go hand in hand: already in 1981, the first alleged death-by-computer-game occurred.[29] The misgivings toward binary thinking expressed earlier by Derrida and then by Deleuze and Guattari at the dawn of the 1980s may strike us as prescient, now that so many of the public debates in our moment center around the beneficial and harmful aspects of life governed by ones and zeros.

Binarization is not the only dangerous aspect of root- or rootlet-centered thinking, in their view. They qualify rootedness as hazardous, a direct rebuttal of the position of Weil and Heidegger, who viewed rootedness as a necessity for human well-being. For Deleuze and Guattari, the root and the rootlet are enemies of desire, of spontaneity, and of multiplicity. Root thinking follows the closed narrative of beginning, middle, and end; rootlet-thinking works in seemingly chaotic but repetitive circles. Both kinds of thought originate in a center. Deleuze and Guattari summarize the characteristics of the rhizome, which embodies for them an alternative to the exhausted and exhausting root and rootlet systems. The rhizome "connects any point to any other point," has no clear beginning nor end, and represents an "anti-genealogy."[30] Deleuze and Guattari want to move away from the classical two-dimensional format of the traced root on paper and expand it into a four-dimensional phenomenon, plastic and dynamic in time and space. Following a logic of accumulation ("and . . . and . . . and . . .") rather than selection ("either/or"), the rhizome

29. An American named Jeff Dailey allegedly had a heart attack after posting a top score on the game *Berzerk*. He was nineteen years old.

30. Deleuze and Guattari, *A Thousand Plateaus*, 21.

accepts anything and everything as a potential line of continuance. Our authors advocate an expansion of the field of possibilities, wondering why must we be limited by cultural impulse or habit. They imply that culture suffers from something akin to repetitive motion disorder; in copying by reflex the same compartmentalizations and hierarchies, we ultimately do harm to culture, causing it to break down. In the more epistemologically oriented portions of the book, they reject an arborescent depiction of the brain and its processes.[31] Instead, they propose a more complex and less hierarchical model composed of "acentered systems."[32] Decentralization is one of the primary characteristics of a rhizomatic epistemology. As Emily Apter writes, "The rhizome functions as a social pathogen, an antistatist virus or Agent Orange of nonassimilation,"[33] an image that conveys the invasive nature of the rhizome in the centered social sphere. It is not that the rhizomatic form does not have what might be called "points of interest," with lines intersecting to form meaningful junctions. The nodal integrity of the botanical form is maintained, but these convergences are multiple and nonhierarchical. Rather than a central taproot through which all energies are dispensed and received, thought is conceived as an erratic pattern sustained in its spontaneous dynamism by acentric stores of energy. Privileging short-term over long-term memory—and it seems that they mean "memory" in both an individual and a collective sense[34]—they argue that the rhizomatic instantaneity of the former operates under the conditions of "discontinuity, rupture, and multiplicity" (16).

The building blocks of the rhizome are described alternately as molecules, atoms, particles, genes, and other microscopic units, but the most interesting of these components are fibrous forms like the strand, the string, and the filament, which Deleuze and Guattari use to create an antigenealogical string theory. They find in the string a more or less elementary form with a tensile nature; it can be elongated, and it may function as a line of connection between two points. In its abstract and geometrical form, they write merely of the line.

31. They write: "Thought is not arborescent, and the brain is not a rooted or ramified matter." Ibid., 15.

32. "To these centered systems, the authors contrast acentered systems, finite networks of automata in which communication runs from any neighbor to any other, the stems or channels do not preexist, and all individuals are interchangeable, defined only by the *state* at a given moment—such that the local operations are coordinated and the final, global result synchronized without a central agency." Ibid., 17.

33. Apter, *Continental Drift*, 31.

34. By their definition, long-term memory involves "family, race, society, or civilization." Deleuze and Guattari, *A Thousand Plateaus*, 16.

They argue that writing, life, and destiny are all made of lines.[35] One of their favored expressions, *lignes de fuite*, or lines of flight, describes the volatile vectors of the rhizome, the wild cards of the form. These lines share with essayistic digression the will to abandon discursive and formal predictability, and they avert the possibility of stable dualisms. The authors introduce the fiber, a small organic unit transferable between humans and animals (and implicitly plants as well), into the rhizomatic vocabulary to illustrate the hyphenated concept of "becoming-(animal, human, etc.)." Because these fibrillary units change place, "dis-integrating" themselves from one body only to join another, they essentially dissolve the secure and insular identity of the individual. As turbulent protuberances, they break away and tack themselves at will to other forms. Like a shag carpet whose fibrils detach and reattach themselves to other things (skin, clothing, pets, furniture), the borders of the body are more fuzzy than that of a block-solid. In Deleuze and Guattari's vision, bodies are filamentary accretions. They also give a strandular form to the elements that make up the rhizomatic structure, using a metaphor borrowed from Ernst Jünger: that of marionette strings. They write, "Puppet strings, as a rhizome or multiplicity, are tied not to the supposed will of an artist or puppeteer but to a multiplicity of nerve fibers, which form another puppet in other dimensions connected to the first" (8). The marionette strings are not pulled by a performer but by the neural dendrites, the live wires of the nervous system that set his body in motion. These are strings that pull strings. The fatherly manager has been ousted and replaced by the billions of strands responsible for all of his agency. This new shift in responsibility actively decenters central power.

And if the father is decentered, the family tree is thrown off-kilter. In their antigenealogical string theory, Deleuze and Guattari replace *filius* (*le fils*, "the son") with (*les fils*, "strings").[36] They accept the fact of influence, but genealogical transmission of influence is where they draw the line. Their new strandular logic does away with the straightness of bloodlines and the neatness of hereditary transmission through filiation. This was already a major theme in *Anti-Oedipus*, which juxtaposed filiation and alliance as social, not biologic,

35. "We are made of lines. We are not only referring to lines of writing. Lines of writing conjugate with other lines, life lines, lines of luck or misfortune, lines productive of the variation of the line of writing itself, lines that are *between the lines* of writing." Ibid., 194.

36. In his *The World, the Text, and the Critic*, Edward Said makes the distinction between filiation, which "belongs to the realms of nature and of 'life,'" and affiliation, which "belongs exclusively to culture and society." Said, *World, Text, and Critic*, 20.

realities.[37] Favoring production over reproduction, they explain in *A Thousand Plateaus* that "a rhizome is not amenable to any structural or generative model" (12). Disregarding the distantly rooted forebears, Deleuze and Guattari invest their energies in the figure of the child, the highest-reaching branch of the family tree, for a way out of rooted thinking. For them, the child manages to break loose from the system of reproduction.[38] In a move contrary to that of thinkers who look toward the ancestors for wisdom, they look to the child as a salvational model. History and heredity are shakable habits; they can be replaced by the improvisational and uninhibited spontaneity of the child. Deleuze and Guattari are tired of the tree, in its genealogical form and in its role as epistemological figure; a way out of arborescent logic is the recognition of the multiple elementary forms that make it, those young and capricious strands that undo the integrity of the tree as a cohesive unit. The intellectual self-orphaning of Deleuze and Guattari is somehow worked through in the nurturing of this metaphorical child, wrested from the oppressive family line but in need of guardianship.[39]

The "becoming-elementary" entailed in this particularization of matter and thought is not far removed from other forms of elementarity we have encountered throughout this study. The various iterations of "becoming-elementary," what could equally be called "becoming-original," "becoming-primeval," "becoming-autochthonous," "becoming-foundational," "becoming-fundamental," and other improvised compounds, entail a scrutiny of the building blocks of existence and of civilization. Deleuze and Guattari have replaced the phenomenologists' emphasis on Being with a new emphasis on Becoming.

37. They write, "It serves no purpose to recall that genealogical filiation is social rather than biological, for it is necessarily biosocial inasmuch as it is inscribed on the cosmic egg of the full body of the earth. It has a mythical origin that is the One, or rather the primitive one-two." Deleuze and Guattari, *Anti-Oedipus*, 154.

38. "In the case of the child, gestural, mimetic, ludic, and other semiotic systems regain their freedom and extricate themselves from the 'tracing.'" Deleuze and Guattari, *A Thousand Plateaus*, 15.

39. In the introduction to the English translation, Brian Massumi writes of Deleuze, "He discovered an orphan line of thinkers who were tied by no direct descendance but were united in their opposition to the State philosophy that would nevertheless accord them minor positions in the canon." Ibid., x. These philosophers included Lucretius, Hume, Spinoza, Nietzsche, and Bergson. Deleuze perhaps includes himself in this family of orphans. Robert Tally Jr. argues that "it is in his earlier works that Deleuze most carefully identifies that nomadic line of flight within the Western philosophical tradition, the counter-history of philosophy or nomadography that typifies Deleuze's radically creative engagement with philosophy." Tally, "Nomadography," 17.

Rather than an appeal for a return to the foundations, they favor probing the composition of these forms in order to show that they are much less solid than imagined. David Macauley, in his book *Elemental Philosophy*, traces a genealogy from Empedocles's ideas on elemental roots to Deleuze and Guattari's rhizomatic thinking. He writes that "Empedocles' ideas may be reasonably interpreted so as to show an anticipation or foreshadowing of the theory of evolution, an attentiveness to and deep sympathy with sentient organisms and nonhuman entities, and a discourse germane to comprehending environmental contamination."[40]

A primary reason for the establishment of this genealogy is Empedocles's use of the term *rhizomata*, which means the roots or elements of earth, water, air, and fire.[41] Empedocles's fifth-century elemental iteration of rootedness is an early instance of the connection between what is *basic* to our cosmos, in terms of both cause and composition, and what burgeons distantly from an original event. When Deleuze and Guattari take up the elemental becoming, they do their best to remain antifoundational, but in a way that differs from Derrida. Derrida points to the quiet chokehold of the ground on all Western thought and seeks to make this thought uproot itself by showing the contradictions, the lacunae, and the violences inherent to it, whereas Deleuze and Guattari pulverize this ground, breaking it into its smallest parts, thus undoing the solidity of what was thought to be *terra firma*. They point to the spaces in between the constitutive units of the foundation, suddenly revealing the looseness of the terrain on which Western thought is positioned.

To bind their compositional metaphors (the layer, the particle, the gene, the assemblage, the string, the element, etc.), Deleuze and Guattari devote a significant amount of attention to the relationships between these composite parts. Their thought is optimized in a four-dimensional framework (three spatial dimensions in time), allowing them to incorporate time and space into their discussions of various relationships between the discrete units mentioned above (superimposition, chronology, simultaneity, analogy, yoking, adjacency). The rhizome, with a firm shape in its nonmetaphorical form, is rendered in their writing as a hazier figure, similar to fog, blur, and other forms

40. Macauley, *Elemental Philosophy*, 104.

41. Macauley argues, "Empedocles is the first philosopher to thematize the four elements as necessary and sufficient agents to explain the entire world order, and by ostensibly using the terminology of roots, he sets the tenor and trajectory of later thinking about them. The language of roots is significant because it continues to function in philosophy in the sense of something elemental, grounded or foundational, including even etymological 'roots.'" Ibid., 72.

of in-between-ness, what they refer to as haecceity.[42] They emphasize the indefiniteness of form and its ability to fool the one who perceives it as uncompromisingly solid. What is refreshing in Deleuze and Guattari's thought, and what accounts for its loud and long resonance, is its permission to take seriously the foggy, wordless, in-between notches in culture.

In their aim to convince, they desperately seek a noninstitutionalized method to reorient the institution, for any method sanctioned by "the system" would be inherently flawed. Philosophy is a discipline composed, in large part, of countless networks of debates, mostly between men, about things of which no one can ever be certain but which must be portrayed as knowable in order for the discipline to continue. Theory, as I understand it, is a synonym for philosophy minus the reputation of being governed by *les éminences grises*, as it publicizes itself as more accommodating to a broader demographic. While we may attribute to philosophy a more arborescent logic and to theory a more rhizomatic one, the "treeness" (*Baumheit*) is no less present in the rhizomatic form. Cutting down the tree of philosophy, pruning it, uprooting it, decorating it, tapping it, climbing it, contemplating it, building a house in it or out of it, or giving it another name: it seems that thought cannot escape arborealization in the Western imagination. A tree is inseverably a tree.

On the other hand, if we approach the coauthored texts of Deleuze and Guattari as literature—given that they call attention repeatedly to their own aesthetic making, they rely on metaphors and vivid imagery, and their experimental character has clear antecedents in literary prose and poetry—there is less pressure to sacralize their thought. Because they are both philosophers, one is tempted to read everything they write as philosophy, but I believe the rhizomatic and nomadic ruminations of Deleuze and Guattari have had their most significant impact not in philosophy, not in politics, not in history, but in poetics. In consonance with Elizabeth Bruss, who analyzed the aesthetic aspects of theory in her book *Beautiful Theories: The Spectacle of Discourse in Contemporary Criticism* (1982), I read Deleuze and Guattari's coauthored books as rich literary texts that unfix the categories of essay, history, and theory. Compounded by the fact that theory, like fiction, does most of its work in the virtual mode, the imaginative component of Deleuze and Guattari's collaborations could qualify them all the more as literary texts. In his *A Nomad Poetics: Essays* (2003), the Luxembourger-American writer Pierre Joris writes, "What

42. "Haecceity, fog, glare. A haecceity has neither beginning nor end, origin nor destination; it is always in the middle. It is not made of points, only of lines. It is a rhizome." Deleuze and Guattari, *A Thousand Plateaus*, 263.

is needed now is a nomadic poetics. Its method will be *rhizomatic*. . . . A nomadic poetics will cross languages, not just translate, but write in all or any of them."[43] He continues, explaining that a nomadic poetics will take "into account not only the manifold of languages & locations but also of selves each one of us is constantly becoming" (44).[44] An outstanding contemporary example of a nomad or rhizome poetics in France can be found in the work of Antoine Volodine, whose transnational panoply of narrative voices emits from the living and the dead through time and space, each of his texts energized by what he calls revolutionary shamanism. Patrick Chamoiseau's *Biblique des derniers gestes* (2002) and Claudio Magris's *Alla cieca* (2005) also display many of the characteristics of nomadic, rhizomatic writing. A poetics of the rhizome will undoubtedly be pursued by many writers, consciously or unconsciously, in this first half of the twenty-first century.

Proponents of Deleuze and Guattari should not be disappointed if their texts are taken for "merely" literary ones; the most staggering cultural changes often must happen first in literature, music, and art, which provide the space where new ideas may be tried out without the requirement of apodictic certainty. Throughout this book, I've made the case that human rootedness is the dangerous literalization of a metaphor. If literature welcomes metaphor and, better yet, invites the abandoning or refashioning of old ones and the creation of new ones, then it is possible to imagine literature as a potential site of the deliteralization or weakening of certain metaphors. An infiltration through images, an aesthetic crafting of a better metaphor than the root, may be the most effective way to convince people that they are almost never truly bound to the land where they dwell.

THE POSTMODERN PLANTATION

The vibrant afterlife of the rhizome, taken up across disciplines and across borders, seems to attest to its timeliness and relevance. For example, the Martinican philosopher and writer Edouard Glissant, in his books *Poétique de la relation* (1990) and *Philosophie de la relation* (2009) especially, elaborated the rhizome concept in his theories of Creolization, archipelagic thinking, *Relation*, *métissage*, the *chaos-monde*, and the new Caribbean iterations of baroque expression. Because of the French Caribbean's complex political relationship

43. Joris, *Nomad Poetics*, 5.

44. His description ends with this: "The nomadic poem as ongoing & open-ended chart of the turbulent fluxes the dispersive nature of our realities make inevitable." Ibid., 44.

with France, the negotiation of identity there showed the shortcomings of the root metaphor, which was never as satisfying as it might have been for *un Français de souche* living in the *métropole*.[45] Throughout the twentieth century, French Caribbean writers and intellectuals took on the project of tracing their roots, first in the 1930s and 1940s through the *Négritude* movement (which located their roots in Africa), then in the 1960s through the *Antillanité* movement (which located their roots in the Caribbean, in the specificities of each island), and in the late 1980s and 1990s through the *Créolité* movement, which claims a centerless, rhizomatic, pan-Atlantic network of relationships between the wide varieties of languages and cultures that find their home on the islands. Glissant writes, "Rhizomatic thought is the principle behind what I call the Poetics of Relation, in which each and every identity is extended through a relationship with the Other."[46] He recognizes the extent to which the rhizome metaphor conforms to the very habits of thought it seeks to subvert but argues that its value lies in its will to complicate and to open up the possibility of new relations. Glissant proceeds with a celebration of errantry and a critique of filiation and its conscious occultation of beginnings in order to preserve the Mystery. He is less prone to think in binaries than Deleuze and Guattari; he names binaries—for example, *nomadisme* and *sédentarité*—and then proceeds to show the great variety within each of these categories, a variety that demolishes the dichotomy.

The Cuban writer Antonio Benítez-Rojo has also adopted the vocabulary of Deleuze and Guattari in perhaps his best-known book, *The Repeating Island: The Caribbean and the Postmodern Perspective* (*La isla que se repite: El Caribe y la perspectiva postmoderna*), published in 1989. Benítez-Rojo replaces the terrestrial with the aquatic, arguing that in the Caribbean, water rather than earth is the necessary departure point for thought. This liquefaction of Deleuze and Guattari's geophilosophy takes shape in Benítez-Rojo's deployment of a natalist and vitalist language of the source, the womb, the mother, and water as the elixir of life. He is acutely aware of the terrestrial instability of the Caribbean, what he calls "the extremely complex and difficult architecture of secret routes, trenches, traps, caves, breathing holes, and underground

45. For more on the distinction between root and rhizome in Caribbean philosophy and linguistics, see Richard L. W. Clarke, "Root versus Rhizome: An 'Epistemological Break' in Francophone Caribbean Thought," *Journal of West Indian Literature* 9, no. 1 (April 2000): 12–41; and Sabine Hofmann, "Against Monolinguism: Roots, Rhizomes and the Conceptualization of Creole Language," *Matatu* 27–28 (2003): 277–85.

46. Glissant, *Poétique*, 23.

rivers that constitute the *rhizome* of the Caribbean psyche."[47] The Caribbean chaos-world, described so elegantly by Glissant, is equally present in Benítez-Rojo's depiction of what goes on below the arborescent model described by Deleuze and Guattari. He writes, "Beneath the turbulences of *árbol, arbre*, tree, etc., there is an island that repeats itself until transforming into a meta-archipelago and reaching the most widely separated transhistorical frontiers of the globe" (24). In his trilingual repetition of "tree" and in pinning "etc." to the end of the list, he gestures toward the total *mezcla* of culture and language, a primary characteristic of Caribbeanness, all the while destabilizing the arborescent model by showing that on the islands, culture never organized itself as a clean, bifurcating structure, the way European culture is often (erroneously) depicted. The archipelago is a productive node that disrupts time and space, undoing national boundaries and fusing various strands of time. He argues that "one of the most clearly and frequently seen regularities of the Caribbean novel is its reiteration of the theme that has come to be known as 'the search for identity' or 'the search for roots'" (186). As I have shown throughout this book, the search for roots seems to be a universal problem. However, the nature of the search for roots in the Caribbean barely resembles the search for roots in, say, Europe. The story of how people ended up in each of these respective places differs too greatly.

Perhaps more important than their shared vocabulary of the rhizome, the chaotic, the baroque, and the archipelago is the emphasis by Glissant and Benítez-Rojo on the importance of the plantation in the formation of Caribbean culture. Deleuze and Guattari depict Westerners as root-seeking forest-dwellers and makers of fields, while Easterners are portrayed as restless occupants of the steppe and makers of gardens.[48] Glissant and Benítez-Rojo focus on one particular kind of European field-making, the colonial planta-

47. Benítez-Rojo, *Repeating Island*, 255.

48. It is odd how the tree has dominated Western reality and all of Western thought, from botany to biology and anatomy, but also gnosiology, theology, ontology, all of philosophy . . . : the root-foundation, *Grund, racine, fondement*. The West has a special relation to the forest, and deforestation; the fields carved from the forest are populated with seed plants produced by cultivation based on species lineages of the arborescent type; animal raising, carried out on fallow fields, selects lineages forming an entire animal arborescence. The East presents a different figure: a relation to the steppe and the garden (or in some cases, the desert and the oasis), rather than forest and field; cultivation of tubers by fragmentation of the individual; a casting aside or bracketing of animal raising, which is confined to closed spaces or pushed out onto the steppes of the nomads. The West: agriculture based on a chosen lineage containing a large number of variable individuals. The East: horticulture based on a small number of individuals derived

tion machine, which is both a violent system and the painful birthplace of a hybridized, global future. What is a plantation? It is a highly organized system that forces the encounter between labor and vegetal life. It is the site of hierarchical power struggles, of botanical and human resource management, and of achieving economic success for one population at the back-breaking expense of another. Because, as I mentioned earlier, plant material is not the ideal surface for lasting inscription, the story of the plantation has largely gone untold. Glissant and Benítez-Rojo hope to change this by showing how the plantation has left its mark on every aspect of Caribbean culture and thought. In his chapter "Lieu clos, parole ouverte," Glissant notes that the plantation system is what links the southern United States, the Caribbean, the Caribbean coast of Latin America, and northeastern Brazil. It is thus a place of Relation. Despite the great cultural differences that divide these regions, the plantation system creates a common structure and thus puts these seemingly disparate places in constant conversation. The pyramid structure of plantation power provided it a certain level of stability, and within its confines—Glissant has termed it a "lieu clos"—the plantation managed to become what he calls "one of the focal points where some of the contemporary modes of Relation were elaborated."[49] Although Glissant hesitates to explicitly define Relation, he gives hints of what it might mean. It is a system of connection. It is *not* totality. It is both an internal and an external movement that becomes an entirely new and unstable thing with each new element added to or taken away from it. It is the "limitless effort of the world" ("l'effort sans limites du monde" [186]), emptied of ideology and interested only in developing infinite points of contact between cultures. It links, relates, and relays. Despite the inhumanity of the plantation, "humanities" (79) managed to thrive there. This agricultural penitentiary created a proliferation of "stories, proverbs, sayings, songs" (82), because the oral tradition could not be stopped by the rolling acres of sugar cane. He argues that the neat and confident filiation presented in Alex Haley's *Roots* does not correspond to "the talking genius of our lands" ("génie parlant de nos pays"), which is much hazier. In Caribbean genealogies, "our origins multiplied" ("nos souches se sont démultipliées" [86]). He calls the plantation "the second matrix" (87) after the slave ship, one that generated a shared

from a wide range of "clones." Does not the East, Oceania in particular, offer something like a rhizomatic model opposed in every respect to the Western model of the tree?

Deleuze and Guattari, *A Thousand Plateaus,* 18

49. "La Plantation est un des lieux focaux où se sont élaborés quelques-uns des modes actuels de la Relation." Glissant, *Poétique*, 79.

literature, language, and music between Bahia, Harlem, Jacmel, and Fort-de-France. In short, the plantation is a laboratory and one of the "wombs of the world" in which spatial closedness translates to cultural openness (89). The global encounter of cultures was tested first on a small scale in the Caribbean, making it a kind of archipelagic avant-garde. In Benítez-Rojo's chapter "From the plantation to the Plantation," he puts the problem in similar terms, insisting on the plantation as both a figurative and a nonfigurative phenomenon. He suggests that one should not begin to "read" the Caribbean through culture but rather through socioeconomic circumstances, and doing so leads him to the plantation as a starting point for his analysis. For him, the plantation "could serve as a telescope for observing the changes and continuities of the Caribbean galaxy through the lenses of multifold disciplines, namely, economics, history, sociology, political science, anthropology, ethnology, demography, as well as through innumerable practices, which range from the commercial to the military, from the religious to the literary."[50] This list resembles the list of disciplines that took up the rhizome as a central concept in the late twentieth and early twenty-first centuries, suggesting that Benítez-Rojo has big ambitions for his notion of the plantation with a capital P, which he uses to distinguish actual individual plantations (small p) from "the society dominated by plantation economy" (317n8). He argues that if plantations had never been established, the populations of the islands would be "miniature replicas—at least in demographic and ethnological terms—of the European nations that colonized them" (39). Thus the plantation was the necessary precondition for the diverse chaos-world of the Caribbean and for the region's "huge and heteroclitic archive" (313) of letters and traditions. The "three great paradigms of knowledge" that he explores in *The Repeating Island* are the People of the Sea, modernity, and postmodernity (314), and the plantation, both literal and figurative, is the primary site where the conjunction of all three occurs, often in bewildering ways. These various encounters in time and space have been described by people of the Caribbean as "syncretism, acculturation, transculturation, assimilation, deculturation, indigenization, creolization, cultural *mestizaje*, cultural *cimmaronaje*, cultural miscegenation, cultural resistance, etc." (37), a fact that leads him to wonder how one can "be sure that a Caribbean culture even exists," given the unmanageable heterogeneity that often causes language to fail. Benítez-Rojo proceeds by using the socioeconomic system of the plantation to read the literature of the Caribbean. One cannot disregard this system, he posits, because every word, every thought issues from it. His

50. Benítez-Rojo, *Repeating Island*, 38.

conclusion: "The plantation turns out to be one of the principal instruments for studying the area, if not indeed the most important" (39). It is clear that in their appropriation of Deleuze and Guattari's metaphor of the rhizome, Glissant and Benítez-Rojo have customized it for the islands. The closed system of the plantation made the Caribbean rhizome possible. In fact, the Caribbean was the only postmodern aspect of modernity. Without the cultural combinatorics that were part and parcel of plantation life, the islands would not have become the laboratory of the future. This kind of rhizomatic thinking takes the tragedy of plantation history and transforms it into a redemptive narrative, turning suffering into possibility. The suggestive concept of the rhizome seems to be fully customizable. The new proponents of vegetal democracy have certainly found use in its application.

NEO-PAGANISM AND PLANT DEMOCRACY

> It can also happen, if will and grace are joined, that as I contemplate the tree I am drawn into a relation, and the tree ceases to be an It.
>
> MARTIN BUBER, *I and Thou*

> This confusion of things for persons, this ontological error of taking an It for a Thou, is exactly what the prophets called paganistic. It is the magical thinking that the philosophers wanted to overcome.
>
> VILÉM FLUSSER, "Taking Up Residence in Homelessness"

Deleuze and Guattari—and Derrida as well—were responsible in part for a fascinating tendency in philosophy and critical theory today; they are, if nothing else, at least to be credited for providing a language of the posthuman. As the refusal of an anthropocentric world, there have been many attempts in recent decades to redistribute attention, care, and even rights to nonhuman and even nonliving entities. The history of the West might be described as the (often failed) attempt to apportion certain resources, privileges, and respect as evenly as possible. Remaining for the moment solely in the realm of the human, there have been hard-fought efforts over the centuries to give rights to the disenfranchised: women, racial or sexual minorities, immigrants, children, the poor, the colonized. The attempt was then extended to animals, most notably through efforts such as Peter Singer's animal liberation movement in the mid-1970s. We are now witnesses of efforts at extending respect to plant life, to entire ecosystems, and to the environment, which has been allegorically transformed into a person named Mother Earth. For example, Michael Marder argues, "Rather than sentience, it is the finitude of a living being that furnishes the 'yardstick' for ethical treatment,"[51] meaning that not just plants but anything with a finite existence deserves an elevated status alongside the human. To take this tendency one step further, several recent books have given a similar attention to things that were never alive at all, such as rocks, television sets, and precious metals.[52] What is next? A respect for what doesn't exist? A respect for imagined or unimaginable worlds?

The attempt to break down the barrier between culture and nature, between *anthropos* and all that isn't *anthropos*, is a signature of our time. We find it, for example, in Philippe Descola's impressive study *Beyond Nature and Culture* (*Par-delà nature et culture*, 2005), in which he shows that this separation is characteristic of a specifically Western and relatively recent ontology. In the preface to his book, he uses an architectural metaphor—that of a house with two floors—to describe the tendency to compartmentalize culture and nature. He argues that "various bizarre beliefs are seeping down to the ground floor" and upsetting this separation. These beliefs include "fragments of Eastern philosophy, remnants of hermetic Gnosticism, or multifaceted New Age systems, none of them very serious but liable . . . to weaken the barriers that have been constructed to separate humans from nonhumans."[53] In this final section, we'll see how such mysticism has recently impacted thought about rootedness.

Since our focus here is the botanical world, I will limit my analysis to the new calls for vegetal democracy, even though it must be acknowledged that this is just one iteration of the tendency to move away from anthropocentrism toward a multicentered conception of the universe. Regarding plant life alone, something is happening that will, some believe, shake the foundations of philosophy and science, something Michael Marder calls the Copernican revolution of botany. Three recent books seek to redefine the relationship between human and plant. Matthew Hall's *Plants as Persons: A Philosophical Botany* (2011), Eduardo Kohn's *How Forests Think: Toward an Anthropology beyond the Human* (2013), and Michael Marder's *Plant-Thinking: A Philosophy of Vegetal Life* (2013) call for a revolution of ecospheric proportions, one that involves an overhaul of the anthropocentric drive to view plants as mere

51. Marder, *Plant-Thinking*, 187.

52. See, for example, Jane Bennett, *Vibrant Matter: A Political Ecology of Things* (Durham, NC: Duke University Press, 2010); Ian Bogost, *Alien Phenomenology; or, What It's Like to Be a Thing* (Minneapolis: University of Minnesota Press, 2012); and Jeffrey Jerome Cohen, ed., *Animal, Vegetable, Mineral: Ethics and Objects* (Washington, DC: Oliphaunt Books, 2012).

53. Descola, *Beyond Nature and Culture*, xvi.

resources for human exploitation. The problem is that this revolution was already attempted once in the 1970s, even before the publication of Singer's faunal call to arms, and it was received mostly as a joke or as an amusing entertainment with no real bearing on human behavior. Peter Tompkins and Christopher Bird, in their 1973 hit book *The Secret Life of Plants*, attempted to shake the foundations of anthropocentric thought by suggesting that plants not only are sentient, but that they may have a cellular consciousness that could be superior to that of humans. The authors look toward a wide variety of sources for evidence of cellular consciousness in plants: the botanical experiments of Sir Jagadish Chandra Bose, Charles Darwin, Vladimir Soloukhin, and others; the spiritual or occult writings of Rudolf Steiner and Gustav Fechner, and others like Goethe, whose thought was influenced by Rosicrucianism; and research in parapsychology, biochemistry, electrophysiology, geomagnetism, and radionics. Despite the book's enormous commercial success as a *New York Times* bestseller, skeptics emerged from the woodwork to critique the authors' shoddy method and their willingness to accept occult pseudoscience and folkloric anecdotes as truth. As a reviewer named Elsa First put it, *The Secret Life of Plants* was "the funniest unintentionally funny book of the year."[54]

The biographies of Tompkins and Bird point toward a passion for all forms of arcane knowledge and the world of secrets. During World War II, Tompkins worked in Italy as an undercover agent for the OSS (Office of Strategic Services) and wrote many occult-themed books, with titles such as *Secrets of the Great Pyramid, The Secret Life of Nature*, and *The Magic of Obelisks*. Bird worked for the CIA in the 1950s and wrote a book called *Divining Hand: The Art of Searching for Water, Oil, Minerals, and Other Natural Resources or Anything Lost, Missing, or Badly Needed*. They coauthored other books that confirmed their infatuation with the underground—in its literal and its figurative sense—including *Secrets of the Soil* and "*Pre-historic Science: Ancient Maps Prove Existence of Advanced Scientific Civilization Four Thousand Years before Christ*," which was left as an unpublished manuscript. The subterranean in all its forms was their obsession. As I have shown throughout this study, the belief in a metaphysical connection between plants and humans is often characterized by an underlying attraction to paganism, to mystery, and to arcane and primordial knowledge. Tompkins and Bird are no exception; they reframe this tendency for an American mass audience. As we have seen, Plato described

54. First, "Secret Life of Plants," 15. She writes, "Peter Tompkins and Christopher Bird have concocted a popular-science pastiche of these New Occult hopes and brought them out into the marketplace, glibly tailored to bid for middle-class respectability" (15).

the human as a plant whose soul was a root embedded in the ether. Aristotle claimed that plants possessed a vegetative soul, also shared by animals and humans, a notion taken up by the Thomists in the Middle Ages. Published in 1848, Gustav Theodor Fechner's book *Nanna, or the Soul-Life of Plants* made the case for the existence of a vital psychic force in the botanical world. Fechner, who accidentally blinded himself during a series of experiments that involved staring directly into the sun, claimed he could see the souls of flowers after recovering his vision. Heidegger argued in the first half of the twentieth century that humans are grounded in their native soil and that just as the soil imparts certain defining characteristics to the flora of a particular region, the human is also marked by its place of origin. The French philosopher Emmanuel Levinas considered Heidegger's obsession with rootedness as a kind of residual paganism, a childish tendency to view the natural world through the lens of mysticism. This begs the question: Why do humans so readily identify with plants? Why do we believe in rootedness? And why are these beliefs so consistently bound up in mystical tropes of buried origins and lost primordial connections?

There have been several other recent notable attempts to reshape our perceptions of plants. Daniel Chamovitz has written a book about what plants know and argues that they do indeed have senses. After an experiment involving plants' response to light, Chamovitz discovered in plants and humans a shared group of genes for determining brightness, a discovery that "led to the obvious question as to what these seemingly 'plant-specific' genes do in people."[55] He is careful to distance himself from Tompkins and Bird, writing, "Worse than leading the unwary reader astray, *The Secret Life of Plants* led to scientific fallout that stymied important research on plant behavior as scientists became wary of any studies that hinted at parallels between animal senses and plant senses" (5). Michael Pollan has suggested, in contrast to narratives of the human exploitation of plants, that the plants have managed to engineer human desire to work in favor of their proliferation. Using as examples the apple, the tulip, cannabis, and the potato, he shows how little difference there is between the oblivious bee, carrying pollen from flower to flower in the service of its reproduction, and the human, who obeys the plant's will by killing off some species so that others may thrive, unaware that it is in some sense the successful species' hitman. His thesis is that "human desires form a part of natural history in the same way the hummingbird's love of red does, or the

55. Chamovitz, *What a Plant Knows*, 3.

ant's taste for the aphid's honeydew."[56] For him, humans are less like plants and more like bees.

In the new articulations of plant consciousness, especially Marder's and Kohn's, mysticism seems at first to be replaced by a poststructuralist logic. The books of Hall, Marder, and Kohn question the self-placement of the human at the center of all things and propose instead a new anthropology that gives plants an elevated moral standing. These studies go beyond the claims of animal-rights activists that human supremacy over the animal world is unjust; the same argument is now extended to the botanical realm. These thinkers use the long history of the analogy between plant and human as evidence of an anthropobotanic symbiosis that could hold the key for the human negotiation of coming centuries of environmental threats, political and territorial disputes, lack of resources, and other predicaments that impact both human and plant habitats. These studies argue essentially that the way humans behave and think cannot be separated from a primordial botanical heritage that our species would rather forget.

Matthew Hall's *Plants as Persons* claims that the nature-culture divide is the cause of many of our misunderstandings about the nonhuman world. He argues that "our general, Western, view of plants as passive resources certainly plays a significant role in our ecological plight."[57] At the time of publication, Hall was a research scientist at the Royal Botanic Garden in Edinburgh, where he used his findings to make the case for a new conception of plants as organic beings with agency. A failure to address the needs of the plant world will ultimately imperil humans, he concludes. While people tend to think empathically about animals in certain contexts, plants remain too alien to our species for this connection to occur. The plant cannot look back at us, which gives us the impression that we are alone when facing it. The objective of Hall's book is to find "the most appropriate behavior toward plants in a time of impending ecological collapse" (4). To this end, he challenges the cultural and philosophical biases passed from one generation to the next that force us to see the botanical world through a repressive lens. Campaigning against anthropocentrism, zoocentrism, and dualistic thinking, Hall seeks to reverse human ignorance about plants.

Eduardo Kohn's *How Forests Think* broaches the same questions but from the perspective of anthropology. The book, clearly influenced by poststructuralist thought, insists on a defamiliarization of the human toward itself.

56. Pollan, *Botany of Desire*, xvii.

57. Hall, *Plants*, 4.

His project seeks to undo "anthropocentric narcissism"[58] and replace it with a more decentered conception of the human-nonhuman connection. Using his extensive fieldwork experience in Ecuador among the people of Ávila, who—in his depiction—share a symbiotic relationship with the forest they inhabit, Kohn illustrates how the forest and the creatures that live there are composed of mutually readable signs. He advocates a new literacy that would allow the human to read nature's signs in a profoundly different way and that would acknowledge that we are being read as well. The spirit of Kohn's book is reminiscent of *I and Thou* (1923), written by the Austrian-born Israeli philosopher Martin Buber. In one passage, Buber describes the experience of contemplating a tree: "It can also happen, if will and grace are joined, that as I contemplate the tree I am drawn into a relation, and the tree ceases to be an It. . . . The tree is no impression, no play of my imagination, no aspect of a mood; it confronts me bodily and has to deal with me as I must deal with it—only differently."[59] In contrast to Francis Ponge, who struggled as he tried to engage presence-to-presence with a mimosa tree,[60] Kohn argues that the human's conceptual capacity to think beyond itself is both a gift and a responsibility; the fact that we can think outside our species and outside our moment into the past and the future guarantees us a unique and significant role as stewards in the service of those beings that surround us. While his work widens its focus beyond the plant to the forest (or any ecosystem) as a whole, he uses the same mystically infused language, writing of afterlives, the dreamworld, and soul blindness. Kohn laments the "disenchantment of the world," Weber's famous formulation, arguing that our mechanistic conception has led us toward an ever more meaningless world.[61] As I will show, this claim is a signature of the contemporary poststructuralist (and the postpoststructuralist) mind-set, as evident in Marder's work as in Kohn's.

The most fascinating of the three, Marder's *Plant-Thinking*, is approached primarily through philosophy, with special reliance on hermeneutic phenomenology, deconstruction, feminist and non-Western thought, and Gianni

58. Kohn, *How Forests Think*, 19.

59. Buber, *I and Thou*, 57–58.

60. Ponge writes, with the objectifying eye of the author at work, "Je ne choisis pas les sujets les plus faciles: voilà pourquoi je choisis le mimosa. Comme c'est un sujet très difficile il faut donc que j'ouvre un cahier. Tout d'abord, il faut noter que le mimosa ne m'inspire pas du tout." Ponge, *Tome premier*, 308. Derrida delighted in this text and included it in his *éloge* to Ponge called *Signéponge*.

61. Kohn, *How Forests Think*, 90.

Vattimo's notion of "weak thought."[62] Dismissing Matthew Hall's consideration of plants as persons,[63] Marder believes that twentieth- and twenty-first-century Continental philosophy, specifically what he calls postmetaphysical philosophy, could be the most likely site of a total transfiguration of our thinking about the botanical world. He takes as his task "to give a new prominence to vegetal life" and "to scrutinize the uncritical assumptions on the basis of which this life has been hitherto explained." The book is organized around these central questions: "How is it possible for us to encounter plants? And how can we maintain and nurture, without fetishizing it, their otherness in the course of this encounter?" (3). His book proliferates outward into a full-on critique of Western metaphysics; he sees plants as the encroaching biomass that can grow over the ruins of this kind of thinking. Regarding rootedness specifically, Marder makes a fascinating argument, one that I will cite at length to show the strange endpoint of a loop I've attempted to trace throughout this project:

> [The plant's spatial rootedness in the soil is] a feature responsible for its coding as the figure of unfreedom. Tragically, occidental thought conflates the most plastic form of existence with the most rigid; not only does this view disregard the ontic exuberance and uncontrollable efflorescence of vegetal life, but it also ignores this life's ontological potentialities, still working themselves out in various guises in animals and human beings—the variations that free it to be otherwise than it is. On the one hand, both colloquial and philosophical

62. In their coauthored forward to Marder's book, Vattimo and Santiago Zabala write, "Weak thought not only follows a logic of resistance, but also promotes a progressive weakening of the strong structures of metaphysics." Vattimo and Zabala, in Marder, *Plant-Thinking*, xii. In chapter 4, I argued that Simone Weil's thought is an application *avant la lettre* of Vattimo's weak thought.

63. Marder writes, "It is neither necessary nor helpful to insist, as certain contemporary commentators do, on the need to attribute to vegetal beings those features, like autonomy or even personhood, philosophers have traditionally considered as respect-worthy. To do so would be to render more refined the violence human thought has never ceased unleashing against these beings, for instance by forcing plants into the mold of appropriative subjectivity." Marder, *Plant-Thinking*, 55. In a note, Marder specifies that he's writing specifically about Matthew Hall's *Plants as Persons*. While Marder may take issue with Hall's anthropomorphization of plants, there is a great amount of overlap between their two projects. Both base many of their arguments on the early treatment of plants in Classic Greek philosophy; both challenge the notion of plant passivity; both look toward the East to find alternate ways of thinking about plants; both qualify the human treatment of plants as violent; and both attempt to rethink how we encounter plants using a wide variety of discourses.

> discourses associate the rooted mode of being with immobility and captivity, but, on the other, the perceived indifference of plants interlaces their freedom with human liberty in the domains of ethics, aesthetics, and religion. Despite their undeniable embeddedness in the environment, plants embody the kind of detachment human beings dream of in their own transcendent aspiration to the other, Beauty, or divinity. (12)

In contrast to the countless examples we've seen throughout this book's pages of writers and philosophers who celebrate rootedness, Marder reads in the human's gaze at the plant a kind of pity for its terrestrial captivity. He argues that "the intelligence of plants is not merely a shadow of human knowing, and their behavior is not a rudimentary form of human conduct." Positing that plants are only marginal on a conceptual level, since "vegetal cellulose is the most common organic compound on Earth,"[64] Marder calls philosophy a sublimated plant-thinking and announces a long-overdue consideration of plants as beings that have much to teach us about thought, time, use of resources, the mind-body divide, and silence. If humans conceive of themselves as attached to the land, this could be owed to their vegetal intuition, just one manifestation of the botanical aspects of ourselves. The roots are the site where death is converted to living matter, leading to what Marder calls "a non-mystified and material 'resurrection.'"[65] The reader may note the tensions in Marder's simultaneous celebration and refusal of a mystical logic. Throughout the book, he insists upon transcendental elements of botanical life but tries to find a demystified vocabulary to accomplish this. If, like Kohn, Marder alludes to the Weberian "disenchantment of the world," he does so in order to set up a discourse of what he calls the "post-metaphysical ontology of vegetal life" or "plant-thinking" (18), which opens up the possibility for a new botanical mysticism emptied of God or anything resembling him. Like Monsieur Teste,

64. Marder, "What Is Plant-Thinking?"

65. Similarly, in a peculiar mediation between the living and the dead, caressing the dead with its roots and obtaining nourishment from them, the plant makes them alive again. Vegetal afterlife, facilitated by the passage, the procession of the dead (including the decomposing parts of the plants themselves), through the roots to the stem and on to the flower, is a non-mystified and material "resurrection," an opportunity for mortal remains to break free from the darkness of the earth. Thanks to the plant, fixed in place by its roots, dead plants, animals, and humans are unmoored from their "resting places"; they travel or migrate, just as in certain non-Western religion souls can find their reincarnation in plants.

Marder, *Plant-Thinking*, 67

Valéry's plantlike "mystic without God,"[66] one may recognize Marder as an author who is enough of a child of the Enlightenment to reject irrational, mystical thought but who recognizes the need to give voice to the enigmatic, in-between spaces of existence that rational thought cannot attend to.

Not one of the authors of these three books mentions the widely known precursor *The Secret Life of Plants*, perhaps in a conscious effort to distance themselves from the weird science it contains. We could perhaps call Hall, Kohn, and Marder "intellectual agnostics," thinkers who want to dissolve dualistic thinking by refusing to accept the binaries of East and West, mysticism and rationality, and human and nonhuman. Particularly Kohn and Marder's books rely on poststructuralist vocabulary,[67] and I think it is fair to say that a substantial amount of Heideggerian paganism is to be found in both books. They attempt to do with anthropology and philosophy what religious thought—Jainism, Buddhism, Hinduism, and animism, for example—has already been doing for centuries. In intellectual circles, to call someone a pagan is to accuse the person of elementary religiosity bordering on superstition. This viewpoint considers religion within a progress-oriented, evolutionary framework, implying that faiths have become more complex and sophisticated in time, perhaps even emptying themselves of the mystical element altogether and becoming pure intellect emptied of spirit. For a twentieth-century thinker to return to the bygone or subsumed tropes of a premonotheistic faith in nature is often perceived as a devolution. But one wonders whether, in the examples I've given of the new arguments favoring human-plant syntheses, there is not a tacit admission that paganism got many things right. Have we come full circle? Are we witnesses to a neopaganism that has found a legitimating discourse to camouflage the intellectual's past derision of mysticism?

One could argue that these new attempts to botanize the human and to humanize the plant are simple literalizations of the pathetic fallacy, that old literary projection of human emotions onto nonhuman entities in nature that so frustrated John Ruskin. They could be the result of an ever-more-palpable, species-wide eco-guilt about our treatment of the planet; a compassionate frenzy to extend human rights to everything in sight: animals, plants, water, and rocks; or mere philosophical improvisations of thinkers who resist the

66. Valéry, *Monsieur Teste*, 31. "Alors j'ai dit à M. l'abbé que mon mari me faisait penser bien souvent à un *mystique sans Dieu*." Valéry, *Œuvres*, 34.

67. Many of these terms didn't necessarily originate in poststructuralist thought but have become the buzzwords of the discipline, including *alterity*, *agency*, *animality*, *"always already,"* *finitude*, *post-*, etc.

frameworks that confine them. Or, the botanical intuitions of these thinkers could be absolutely spot on. Whatever the case may be, the near simultaneity of their arrival indicates a sudden consciousness of this particular kind of non-human presence and an impulse to make sense of this awareness. Paradoxically, the fact that we are able to entertain such thoughts reaffirms our absolute difference from animals and plants, who would never be able to have such a thought. By articulating a hope that humans still have the capacity to recontextualize themselves, restoring lost ties with the natural world, we only confirm the impossibility of this task. As in Denise Levertov's poem "Re-Rooting," in which our attempts to "put the roots back" only show the extent of our existential awkwardness, there is nothing less plantlike than the attempt to discover something plantlike in ourselves. It would never occur to a plant to embark on such a venture. The plant is very good at simply *being*, something the human hasn't been good at for millennia.

These new developments allow the rationalist, particularly the atheist or agnostic kind, to entertain the possibility of an afterlife once more. If the root is the place where what has decomposed gets siphoned up again into new life, this means its function is one of resurrection. God is not needed in this life-for-death transaction, or at least he remains behind the scenes. It is possible again to shove away death with the comforting thought that our expired bodies may be upcycled into a Douglas fir.

*

I have argued throughout this book that the most important developments in Continental philosophy of the twentieth-century have involved taking a position on the problem of rootedness, making it one of the most central philosophical concerns from the beginning of the twentieth century to the present. For example, psychoanalysis and the philosophical schools that borrowed from and informed it relied on the exploration of the subterranean aspects of the psyche, the dark symmetry of conscious thinking that remained below ground. Phenomenology depended on a rerooting of philosophy in subjective experience, while existentialism concerned itself with the rootedness of being. Conceptual schools of thought that took interest in the absurd described it as a cutting off of the human from the physical and metaphysical roots that at one time managed to bind people to the world. The obsession with etymological roots touched various schools of thought throughout the twentieth century, particularly among those thinkers we might call "authenticity seekers." As

Ihab Hassan, Emily Apter, and others have argued regarding the shift from a modern to a postmodern sensibility, thinkers, writers, artists, and other cultural agents moved progressively away from a root-based conception of the world to a rhizomatic one. Poststructuralists like Deleuze and Guattari found the rhizome to be helpful in altering the ways we think about and organize knowledge. And the interest of ecologically oriented philosophers, such as Arne Næss and his notion of deep ecology, requires a reconceptualization of the human's relationship with its dwelling place and a consideration of the intrinsic value of the not-human. Finally, the recent calls by thinkers such as Hall, Marder, and Kohn for a more democratic treatment of plant life continue the work begun in poststructuralist circles, which sought a more reciprocally rooted relationship between the human and the nonhuman world.

What is common to all of these late negotiations of rootedness and rootlessness—Deleuze and Guattari's rhizome and nomad, Derrida's *destinerrance*, Glissant's *errance enracinée*, Charles's *enracinerrance*, Flusser's taking up residence in homelessness, Hall's plants as persons, Kohn's thinking forest, and Marder's plant-thinking—is the synthesis of binaries and the categorical refusal to decide. In contrast to the obstinate declarations of someone like Paul Claudel or Maurice Barrès, who were certain of their roots in faith or fatherland, the new roots are just barely roots, always a hybrid of root and something else. It's almost as if we find it impossible to let go of the idea of rootedness but we can't come to terms with its control over us. How is one to refuse an age-old metaphor? By literalizing it? Celebrating its opposite? Changing its name? Hybridizing it? Even those who are opposed to the idea that people have roots tend to celebrate nomadism as its own kind of rootedness. The disconnectedness of the world has seeped into philosophy, which inevitably reflects the spirit of the age. It seems that a real sense of rootedness is only possible for those of a conservative bent, and even then, these loud but timorous claims to a heritage or a place are responses to the threat that this heritage or place is in the process of dissolving.

There seems to be a general intuition in Europe that culture is in decline and the appeal to rootedness is a clear symptom of it. Nature maintains a steady indifference toward cultural peaks and troughs. Perhaps the desire to become rooted in nature is the desire to sidestep the instability of culture's fluctuations and the mercurial aspect of the human in favor of a system that, while dynamic and cyclical, seems to follow a calmer logic. The giant sequoias evoked at the beginning of this book are unconcerned with the rise and fall of democracies. Karl Jaspers, in his essay "A New Humanism," claims that "the disintegration of historic memory is the result of the anti-historic trends in modern technology and politics."[68] The human is essentially at war with its own memory, which it seeks to eradicate. Maybe this is how people want to become more like nature: by nullifying memory, that faculty that distinguishes us from the rest of the living world. Perhaps the human has discovered that memory is too big a burden, that it makes us behave in destructive ways and contributes to our self-extermination. But Jaspers believes that only through memory can the human achieve its full potential. He asks: "Can man break with history? Can he sever his own roots? Can he unfold out of a historic void, and can his mere biologic nature enable him to realize a humanity based on nothing but this biology, this nature, this instant, and seeing and seeking nothing beyond an imaginary charismatic future? No. Man must know what he was, to realize what he can be" (81). Instead of cultivating ties with history, it seems that we've accepted our fate of rootlessness and made the decision not to suffer it alone: instead, we've opted to uproot the whole world along with us, to give ourselves company in this state of uprootedness. The German poet and doctor Gottfried Benn found it difficult to imagine "a God who created anything as gentle as plants and trees" when considering the technological developments of his age: "electronic brains, artificial insemination for cows and women, chicken farms with music laid on to increase productivity, artificial doubling of the chromosomes bringing about giant hybrids, deep freezing, over-heating."[69] Developers of these technologies—meant to make life easier, to smooth time and space, and to open up new possibilities—have forgotten to take something essential into account: namely, those long-term effects that permanently restructure the way the human participates in the world. As specialists, they are trained as micro- rather than macroscopic thinkers, which means they think in terms of parts over wholes. To specialize is to decontextualize. This specialization is the response to a simple fact: The world and what is knowable have become too big for the brain to handle. We favor a more manageable smallness over unwieldy largeness and thus break civilization down into its composites. This cultural contraction has led us to think inwardly, to fold in on ourselves, to celebrate imagined microcommunities, to atomize knowledge, and to move from the big to the small picture. A narcissist is a specialist of the self. We have asked for a compacting force to process the immensity of culture into bite-sized pieces, to summarize and synthesize significant events to save us the time of thinking through them holistically. But by doing so, by caring only about what is im-

68. Jaspers, *Existentialism*, 80.
69. Benn, Primal Vision, 202–3.

mediate and summarizable, we forfeit our own context. If we continue to insist on the root metaphor, it is merely a utopian projection of a more integrated life that seems for the moment unattainable. It could be a matricial memory, similar to the one that draws us to the shores of oceans whence we crawled. We have been out of context for a very long time. The metaphor follows us around to remind us.

Why do some subfields of philosophy and linguistics insist on the importance of context? It is because without it, meaning breaks down. Recall that this is the definition Julia Kristeva gave to abjection, namely the site where meaning breaks down when the separation between subject and object begins to deteriorate. In my reflections on rootedness, a constellation of terms kept appearing: decontextualization, breakdown of meaning, increased abstraction, alienation. It was Simone Weil's writing that nudged me toward a significant conclusion: When people seek roots, they are refusing death's abstraction. The root, in its constant proximity to death, possesses a more than concrete molecular understanding of the process of death's conversion into life. To have roots is to facilitate continuity, to embody the cyclical processes of the natural world, and to disregard the abstract qualities of death by simply making new life.

In recent years, the Swedish company Promessa Organic Burial has developed a method called "promession" for turning a human corpse into compost that can feed a memorial tree or bush above the buried body. After the cadaver is freeze-dried and pulverized, it is buried in a coffin made of corn starch, which decomposes naturally without polluting the surrounding soil. The deceased may opt to have a tree or bush planted above the burial site and fed by his or her body. Like Ovid's tale of the mulberry tree that changed from white to red as the blood of the lovers Pyramus and Thisbe fed its roots and leaves, promession allows the human body to surrender its life essence to another entity that will make good, visible use of it. This accelerated "vegetal afterlife,"[70] as Marder would call it, involves the conversion of dead cells into living ones through the root's transformative proficiency. The body, entrusted to Mother Nature, is guaranteed to live again, a promise that God, in the eyes of skeptics, cannot keep. Another justification for promession is its ethical soundness as an environmentally responsible way to inhume the human. If one cannot treat the environment with care in life, at least a full surrender to it in death—through giving the self over willingly to its decompositional processes—feels like a small compensation. If there is something appealing about this fusion

70. Marder, *Plant-Thinking*, 67.

with nature, it is that such a fusion has become virtually impossible in life. In our mortal awkwardness, we are, it seems, incapable of staying in context. As we leave the rhizosphere behind, harboring a deep nostalgia for it, there is consolation to be found in the fact that it is possible in death to be enfolded in the very constitution of the root.

BIBLIOGRAPHY

Abram, David. *The Spell of the Sensuous: Perception and Language in a More-Than-Human World*. New York: Vintage Books, 1996.

Ahluwalia, Pal. *Out of Africa: Post-structuralism's Colonial Roots*. London: Routledge, 2010.

Alinei, Mario. "Etymography and Etymothesis as Subfields of Etymology." *Folia Linguistica* 16 (January 1982): 41–56.

———. "Thirty-Five Definitions of Etymology; or, Etymology Revisited." In *On Languages and Language*, edited by Werner Winter, 1–26. Berlin: De Gruyter, 1994.

Allen, Donald, ed. *The New American Poetry, 1945–1960*. Berkeley: University of California Press, 1999.

Ansel, Yves. *"La nausée" de Jean-Paul Sartre*. Paris: Editions Pédagogie Moderne, 1982.

Anttila, Raimo. *Historical and Comparative Linguistics*. Amsterdam Studies in the Theory and History of Linguistic Science, series 4, Current Issues in Linguistic Theory. 2nd rev. ed. Philadelphia: Benjamins, 1989.

Apollinaire, Guillaume. *Calligrammes*. Paris: Gallimard, 1966.

Apter, Emily. *Continental Drift: From National Characters to Virtual Subjects*. Chicago: University of Chicago Press, 1999.

Arber, Agnes. *The Natural Philosophy of Plant Form*. Cambridge: Cambridge University Press, 1950.

Arendt, Hannah. *The Jewish Writings*. New York: Schocken Books, 2007.

———. *Responsibility and Judgment*. New York: Schocken Books, 2003.

Aristotle. *On the Parts of Animals*. Translated by William Ogle. London: Kegan Paul & Trench, 1882.

Assouly, Olivier. *Les nourritures nostalgiques: Essai sur le mythe du terroir*. Arles: Actes Sud, 2004.

Atlan, Henri. *Entre le cristal et la fumée: Essai sur l'organisation du vivant*. Paris: Seuil, 1979.

Attridge, Derek. *Peculiar Language: Literature as Difference from the Renaissance to James Joyce*. Ithaca, NY: Cornell University Press, 1988.

Bachelard, Gaston. *La terre et les rêveries du repos: Essai sur les images de l'intimité*. Paris: José Corti, 1948.

Bambach, Charles. "Heidegger, Technology, and Homeland." *Germanic Review* 78, no. 4 (Fall 2003): 267–82.

———. *Heidegger's Roots: Nietzsche, National Socialism, and the Greeks*. Ithaca, NY: Cornell University Press, 2003.

Barrès, Maurice. *Les déracinés*. Vol. 1. Paris: Plon, 1947.

Barthes, Roland. *The Rustle of Language*. Translated by Richard Howard. New York: Farrar, Straus & Giroux, 1986.

Bataille, Georges. "La victoire militaire et la banqueroute de la morale qui maudit." *Critique* 5, no. 38 (July 1949): 789–803.

———. *Visions of Excess: Selected Writings, 1927–1939*. Translated by Allan Stoekl. Minneapolis: University of Minnesota Press, 1985.

Bateson, Gregory. *Steps to an Ecology of Mind*. Chicago: University of Chicago Press, 2000.

Beauvoir, Simone de. *Adieux: A Farewell to Sartre*. Translated by Patrick O'Brian. New York: Pantheon Books, 1984.

———. *The Prime of Life*. Translated by Peter Green. Cleveland: World, 1962.

Beckett, Samuel. *The Letters of Samuel Beckett. Vol. 2, 1941–1956*. Cambridge: Cambridge University Press, 2011.

———. *Waiting for Godot*. New York: Grove Press, 1954.

Bell, Richard H. Introduction to *Simone Weil's Philosophy of Culture: Readings toward a Divine Humanity*. Edited by Richard H. Bell. Cambridge: Cambridge University Press, 1993.

Bémol, Maurice. *Paul Valéry*. Clermont-Ferrand: G. de Bussac, 1949.

Benítez-Rojo, Antonio. *The Repeating Island*. Durham, NC: Duke University Press, 1996.

Benn, Gottfried. *Primal Vision: Selected Writings of Gottfried Benn*. London: Marion Boyars, 1976.

Bennington, Geoffrey, and Jacques Derrida. *Jacques Derrida*. Translated by Geoffrey Bennington. Chicago: University of Chicago Press, 1993.

Berg, Nicolas. *Luftmenschen: Zur Geschichte einer Metapher*. Göttingen: Vandenhoeck and Ruprecht, 2008.

Blanchot, Maurice. *The Infinite Conversation*. Translated by Susan Hanson. Minneapolis: University of Minnesota Press, 1993.

———. *The Writing of the Disaster*. Translated by Ann Smock. Lincoln: University of Nebraska Press, 1986.

Bloch, Ernst. *Heritage of Our Times*. Translated by Neville Plaice and Stephen Plaice. Cambridge: Polity Press, 1991.

Bloch, Howard R. *Etymologies and Genealogies: A Literary Anthropology of the French Middle Ages*. Chicago: University of Chicago Press, 1983.

Blossfeldt, Karl. *Art Forms in Nature*. London: A. Zwemmer, 1929.

Blumenberg, Hans. *Paradigms for a Metaphorology*. Translated by Robert Savage. Ithaca, NY: Cornell University Press, 2010.

Bonta, Mark. "Rhizome of Boehme and Deleuze: Esoteric Precursors of the God of Complexity." *SubStance* 39, no. 1 (2010): 62–75.
Boudon, Brigitte. *Symbolisme de l'arbre*. Paris: Huitième Jour, 2010.
Brosses, Charles de. *Traité de la formation des langues et des principes physiques de l'étymologie*. Vol. 2. Paris: Terrelonge, 1800.
Brown, Charles S., and Ted Toadvine. *Eco-Phenomenology: Back to the Earth Itself*. Albany: State University of New York Press, 2003.
Buber, Martin. *I and Thou*. Translated by Walter Kaufmann. New York: Simon & Schuster, 1970.
Bulwer, Henry Lytton. *The Life of Henry John Temple, Viscount Palmerston with Selections from His Diaries and Correspondence*. Vol. 2. Philadelphia: J. B. Lippincott, 1871.
Burbank, Luther. *The Training of the Human Plant*. New York: Century, 1907.
Camus, Albert. *Œuvres complètes*. Vol. 4. Paris: Gallimard and Club de l'Honnête Homme, 1983.
Carey, Frances. *The Tree: Myth and Meaning*. Burlington, VT: Lund Humphries, 2012.
Cayrol, Jean. *Les mots sont aussi des demeures*. Paris: Seuil, 1952.
Celan, Paul. *Gedichte: Erste Band*. Frankfurt am Main: Suhrkamp, 1975.
———. *Paul Celan: Selections*. Edited by Pierre Joris. Berkeley: University of California Press, 2005.
Celan, Paul, and Nelly Sachs. *Briefwechsel*. Edited by Barbara Wiedemann. Frankfurt am Main: Suhrkamp Verlag, 1993.
———. *Correspondence*. Translated by Christopher Clark. Riverdale-on-Hudson, NY: Sheep Meadow Press, 1995.
Céline, Louis-Ferdinand. *Journey to the End of the Night*. Translated by Ralph Manheim. New York: New Directions, 2006.
Césaire, Aimé. *Cahier d'un retour au pays natal*. 2nd ed. Edited by Abiola Irele. Columbus: Ohio State University Press, 2000.
Chamovitz, Daniel. *What a Plant Knows: A Field Guide to the Senses*. New York: Scientific American/Farrar, Straus & Giroux, 2012.
Charles, Jean-Claude. *Le corps noir*. Paris: Hachette, 1980.
———. "L'enracinerrance." *Boutures* 1, no. 4 (March–August 2001): 37–41.
Cixous, Hélène. *Entre l'écriture*. Paris: Des Femmes, 1986.
———. *Hemlock*. Cambridge: Polity Press, 2011.
———. *Portrait of Jacques Derrida as a Young Jewish Saint*. Translated by Beverley Bie Brahic. New York: Columbia University Press, 2004.
———. *Le voyage de la racine Alechinsky*. Paris: Galilée, 2012.
Claudel, Paul. *L'art poétique*. 15th ed. Paris: Mercure de France, 1943.
———. *A Poet before the Cross*. Translated by Wallace Fowlie. Chicago: Henry Regnery, 1958.
Condé, Maryse. *Crossing the Mangrove*. Translated by Richard Philcox. New York: Anchor Books, 1995.
Darwin, Charles. *On the Origin of the Species, or the Preservation of Favoured Races in the Struggle for Life*. 5th ed. London: John Murray, 1869.

Del Bello, Davide. *Forgotten Paths: Etymology and the Allegorical Mindset*. Bergamo, Italy: Bergamo University Press, 2005.

Deleuze, Gilles, and Félix Guattari. *Anti-Oedipus: Capitalism and Schizophrenia*. Translated by Robert Hurley, Mark Seem, and Helen R. Lane. Minneapolis: University of Minnesota Press, 1983.

———. *A Thousand Plateaus: Capitalism and Schizophrenia 2*. Translated by Brian Massumi. Minneapolis: University of Minnesota Press, 1987.

———. *What Is Philosophy?* Translated by Hugh Tomlinson and Graham Burchell. New York: Columbia University Press, 1994.

Del Vasto, Lanza. *Les étymologies imaginaires*. Paris: Editions Denoël, 1985.

Derrida, Jacques. *La carte postale de Socrate à Freud et au-delà*. Paris: Flammarion, 1980.

———. "Cogito et histoire de la folie." *Revue de la Métaphysique et de Morale* 68, no. 4 (October–December 1963): 460–94.

———. *Dissemination*. Translated by Barbara Johnson. London: Athlone Press, 1981.

———. "Khôra." Translated by Ian McLeod. In *On the Name*, edited by Thomas Dutoit, 89–128. Stanford, CA: Stanford University Press, 1995.

———. *The Margins of Philosophy*. Translated by Alan Bass. Chicago: University of Chicago Press, 1982.

———. *Negotiations: Interventions and Interviews, 1971–2001*. Translated by Elizabeth Rottenberg. Stanford, CA: Stanford University Press, 2002.

———. "Of an Apocalyptic Tone Newly Adopted in Philosophy." In *Derrida and Negative Theology*, edited by Harold Coward and Toby Foshay, 25–71. Albany: State University of New York Press, 1992.

———. *Of Grammatology*. Translated by Gayatri Chakravorty Spivak. Baltimore: Johns Hopkins University Press, 1997.

———. "Les sources de Valéry: Qual, Quelle," *Modern Language Notes* 87, no. 4 (May 1972): 563–99.

———. *Sovereignties in Question: The Poetics of Paul Celan*. Edited by Thomas Dutoit and Outi Pasanen. Bronx, NY: Fordham University Press, 2005.

———. *Sur Parole: Instantanés philosophiques*. La Tour-d'Aigues: Editions de l'Aube, 1999.

———. *Voice and Phenomenon: Introduction to the Problem of the Sign in Husserl's Phenomenology*. Evanston, IL: Northwestern University Press, 2011.

———. *Writing and Difference*. Translated by Alan Bass. London: Routledge, 2001.

Descartes, René. *Lettre-préface des principes de la philosophie*. Paris: Nathan, 1998.

———. *Les principes de la philosophie: Première partie*. Paris: Delagrave, 1855.

Descola, Philippe. *Beyond Nature and Culture*. Translated by Janet Lloyd. Chicago: University of Chicago Press, 2013.

Diderot, Denis, and Pierre Daubenton. "Arbre." In *Encyclopédie, ou dictionnaire raisonné des sciences, des arts et des métiers*. Paris, 1751.

Dumas, Robert. *Le traité de l'arbre: Essai d'une philosophie occidentale*. Arles, France: Actes Sud, 2002.

Esslin, Martin. *The Theater of the Absurd*. New York: Doubleday, 1961.

Felstiner, John. "Mother Tongue, Holy Tongue: On Translating and Not Translating Paul Celan." *Comparative Literature* 38, no. 2 (Spring 1986): 113–36.

———. *Paul Celan: Poet, Survivor, Jew*. New Haven, CT: Yale University Press, 1995.

Ferdière, Gaston. "Vers une spéléopsychologie du verbe." *Revue de Métaphysique et de Morale* 65, no. 2 (April–June 1960): 163–74.

Feser, Edward, and Steven Postrel. "Reality Principles: An Interview with John R. Searle." *Reason,* February 2000, 42–50.

First, Elsa. "The Secret Life of Plants" (review). *New York Review of Books,* December 30, 1973, 15.

Fischer, Sally. "Social Ecology and the Flesh: Merleau-Ponty, Irigaray, and the Ecocommunitarian Politics." In *Merleau-Ponty and Environmental Philosophy*, edited by Suzanne L. Cataldi and William S. Hamrick, 203–15. Albany: State University of New York Press, 2007.

Flaxman, Gregory. *Gilles Deleuze and the Fabulation of Philosophy*. Minneapolis: University of Minnesota Press, 2012.

Flusser, Vilém. "The Cedar in the Park." In *Natural: Mind*, translated by Rodrigo Maltez Novaes. Minneapolis: Univocal, 2013.

———. "Exile and Creativity." In *Writings*, edited by Andreas Stöhl, translated by Erik Eisel, 104–9. Minneapolis: University of Minnesota Press, 2002.

———. "Taking Up Residence in Homelessness." In *Writings*, edited by Andreas Stöhl, translated by Erik Eisel, 91–103. Minneapolis: University of Minnesota Press, 2002.

Foucault, Michel. *The History of Madness*. Translated by Jonathan Murphy and Jean Khalfa. London: Routledge, 2006.

———. "Nietzsche, Genealogy, History." In *Language, Counter-Memory, Practice: Selected Essays and Interviews*, edited by D. F. Bouchard, 76–100. Ithaca, NY: Cornell University Press, 1977.

———. *The Order of Things: An Archaeology of the Human Sciences*. New York: Vintage, 1994.

Frères de l'Instruction Chrétienne. *Histoire d'Haïti*. Port-au-Prince, Haiti: Saint-Joseph, Archevêque, 1942.

Freud, Sigmund. "'The Antithetical Sense of Primal Words': A Review of a Pamphlet by Karl Abel, *Über den Gegensinn der Urworte*, 1884." Translated by M. N. Searl. In *On Creativity and the Unconscious*, 55–63. New York: Harper & Row, 1958.

———. *Civilization and Its Discontents*. Translated by James Strachey. New York: Norton, 1962.

———. *The Interpretation of Dreams*. Hertfordshire, UK: Wordsworth, 1997.

Froidefond, Christian. "Linéarité, verticalité, rectitude: Remarques sur l'*orthotès* platonicienne." *Les Etudes philosophiques* 3 (July–September 1982): 257–79.

Gary, Romain. *The Roots of Heaven*. Translated by Jonathan Griffin. New York: Simon & Schuster, 1958.

Genette, Gérard. *Mimologics*. Translated by Thaïs E. Morgan. Lincoln: University of Nebraska, 1995.

Gide, André. *Essais critiques*. Edited by Pierre Masson. Paris: Gallimard, 1999.

Girard, René. *Deceit, Desire, and the Novel: Self and Other in Literary Structure*. Translated by Yvonne Freccero. Baltimore: Johns Hopkins University Press, 1965.

Glissant, Edouard. *Poétique de la relation*. Paris: Gallimard, 1990.

Gorer, Geoffrey. *The American People: A Study in National Character*. New York: Norton, 1964.

Gourmont, Rémy de. "Les transplantés." In *Promenades littéraires*, 330–47. Paris: Mercure de France, 1919.

Grimard, Edouard. *La plante botanique simplifiée*. Paris: J. Hetzel, 1865.

Guillevic, Eugène. *Un brin d'herbe, après tout: Entretiens avec Jean-Yves Erhel, 21 janvier–28 janvier 1979*. Paris: La Part commune, 1998.

———. *Guillevic: Selected Poems*. Translated by Denise Levertov. New York: New Directions, 1968.

———. *Living in Poetry: Interviews with Guillevic*. Translated by Maureen Smith. Dublin: Dedalus Press, 1999.

———. *Racines*. Illustrated by Robert Blanchet. Boulogne-sur-Seine: Robert Blanchet, 1973.

———. *Relier: Poèmes 1938–1996*. Paris: Gallimard, 2007.

———. *Terraqué*. Paris: Gallimard, 1942.

Guillevic, Eugène, and Raymond Jean. *Choses parlées: Entretiens*. Seyssel, France: Editions du Champ Vallon, 1982.

Guyard, Marie-François. "Claudel et l'étymologie." *Cahiers de l'Association internationale des études françaises* 11 (1959): 286–300.

Hadot, Pierre-Henri. "L'homme, 'plante céleste.'" *Les Etudes philosophiques* 16, no. 3 (July–September 1961): 79–83.

Hägglund, Martin. *Radical Atheism: Derrida and the Time of Life*. Stanford, CA: Stanford University Press, 2008.

Halda, Bernard. *L'évolution spirituelle de Simone Weil*. Paris: Beauchesne, 1964.

Hall, Matthew. *Plants as Persons: A Philosophical Botany*. Albany: State University of New York Press, 2011.

Hamacher, Werner. "The Second of Inversion: Movements of a Figure through Celan's Poetry." In *Premises: Essays on Philosophy and Literature from Kant to Celan*, translated by Peter Fenves, 337–87. Cambridge, MA: Harvard University Press, 1996.

Hammerschlag, Sarah. *The Figural Jew: Politics and Identity in Postwar French Thought*. Chicago: University of Chicago Press, 2010.

Hanania, Cécile. *Roland Barthes et l'étymologie*. Brussels: P.I.E. Peter Lang, 2010.

Harrison, Robert. "America: The Struggle to Be Reborn." *New York Review of Books* 59, no. 16 (October 25, 2012): 64–68.

———. *The Dominion of the Dead*. Chicago: University of Chicago Press, 2003.

———. *Forests: The Shadow of Civilization*. Chicago: University of Chicago Press, 1992.

Hassan, Ihab. *The Dismemberment of Orpheus: Toward a Postmodern Literature*. 2nd ed. Madison: University of Wisconsin Press, 1982.

Heidegger, Martin. *Contributions to Philosophy*. Translated by Parvis Emad and Kenneth Maly. Bloomington: Indiana University Press, 1999.

———. *Gesamtausgabe: I. Abteilung: Veröffentlichte Schriften 1910–1976. Vol. 12, Unterwegs zur Sprache*. Frankfurt am Main: Klostermann, 1985.

———. *Gesamtausgabe 96*. Vol. 3. Edited by Peter Trawny. Frankfurt am Main: Klostermann, 2014.

———. "Memorial Address." In *Discourse on Thinking*, translated by John M. Anderson and E. Hans Freund, 43–57. New York: Harper & Row, 1966.

———. *On the Way to Language*. Translated by Peter D. Hertz. New York: Harper & Row, 1982.

———. *Pathmarks*. Edited by William McNeill. Cambridge: Cambridge University Press, 1998.

———. *Reden und andere Zeugnisse eines Lebensweges, 1910–1976*. Frankfurt am Main: Klostermann, 2000.

———. "The Way Back into the Ground of Metaphysics." Translated by Walter Kaufmann. In *Philosophy in the Twentieth Century*, edited by William Barrett and Henry D. Aiken, 195–212. New York: Random House, 1962.

Hitler, Adolf. *Mein Kampf*. Lavargne, TN: Bottom of the Hill, 2010.

Hollenberg, Donna Krolik. *A Poet's Revolution: The Life of Denise Levertov*. Berkeley: University of California Press, 2013.

Hugo, Victor. *Correspondance, 1815–1835*. Paris: Calmann Lévy, 1896.

Husserl, Edmund. *Cartesian Meditations*. Translated by Dorion Cairns. The Hague: Nijhoff, 1960.

———. *Ideas: General Introduction to Pure Phenomenology*. Translated by W. R. Boyce Gibson. London: G. Allen & Unwin, 1952.

Isidore of Seville. *The Etymologies of Isidore of Seville*. Translated by Stephen A. Barney, W. J. Lewis, J. A. Beach, and Oliver Berghof. Cambridge: Cambridge University Press, 2006.

Jabès, Edmond. *The Book of Questions*. Vol. 1. Translated by Rosmarie Waldrop. Hanover, NH: Wesleyan University Press, 1991.

———. *Le livre des marges I: Ça suit son cours*. Saint Clément de rivière: Fata Morgana, 1975.

Jarrety, Michel. *Paul Valéry*. Paris: Fayard, 2008.

Jaspers, Karl. *Existentialism and Humanism: Three Essays*. Translated by E. B. Ashton. New York: R. F. Moore, 1952.

———. *The Future of Mankind*. Translated by E. B. Ashton. Chicago: University of Chicago Press, 1961.

———. *The Question of German Guilt*. Translated by E. B. Ashton. Bronx, NY: Fordham University Press, 2001.

———. *Tragedy Is Not Enough*. Translated by Harald A. T. Reiche, Harry T. Moore, and Karl W. Deutsch. Hamden, CT: Archon Books, 1969.

Jay, Martin. "Still Waiting to Hear from Derrida." *Salmagundi* 150–51 (Spring–Summer 2006): 25–35.

Joris, Pierre. *A Nomad Poetics: Essays*. Middletown, CT: Wesleyan University Press, 2003.

Jung, C. G. *The Collected Works of C.G. Jung*. Vol. 9, part 2. Translated by Gerhard Adler and R. F. C. Hull. Princeton, NJ: Princeton University Press, 1959.

———. *The Earth Has a Soul: C.G. Jung on Nature, Technology, and Modern Life*. Edited by Meredith Sabini. Berkeley, CA: North Atlantic Books, 2008.

———. "The Philosophical Tree." In *The Collected Works of C.G. Jung*, 13:251–349. Princeton, NJ: Princeton University Press, 1967.

Kirkup, James. "Obituary: Eugene Guillevic." *London Independent*, March 25, 1997, www.independent.co.uk/news/obituaries/obituary-eugene-guillevic-1274938.html.
Kligerman, Eric. *Sites of the Uncanny: Paul Celan, Specularity, and the Visual Arts*. Berlin: Walter de Gruyter, 2007.
Kogl, Alexandra. *Strange Places: The Political Potentials and Perils of Everyday Spaces*. Lanham, MD: Lexington, 2008.
Kohn, Eduardo. *How Forests Think: Toward an Anthropology beyond the Human*. Berkeley: University of California Press, 2013.
Kramer, Matt. "The Notion of *Terroir*." In *Wine and Philosophy: A Symposium on Thinking and Drinking*, edited by Fritz Allhoff, 225–34. Oxford: Blackwell, 2008.
Kristeva, Julia. *Powers of Horror: An Essay on Abjection*. Translated by Leon S. Roudiez. New York: Columbia University Press, 1982.
Kundera, Milan. *Le rideau*. Paris: Gallimard, 2005.
Lacan, Jacques. *The Ethics of Psychoanalysis, 1959–1960*. Translated by Dennis Porter. New York: Norton, 1992.
Lacoue-Labarthe, Philippe. *Poetry as Experience*. Translated by Andrea Tarnowski. Stanford, CA: Stanford University Press, 1999.
La Mettrie, Julien Offray de. *Œuvres philosophiques*. Vol. 2. Paris: Tutot, 1796.
Larousse, Pierre, ed. *Grand dictionnaire universel du XIX siècle*. Vol. 14. Paris: Administration du Grand dictionnaire universel, 1875.
Leiris, Michel. *Glossaire j'y serre mes gloses*. In *Mots sans mémoire* (Paris: Gallimard, 1969).
Levertov, Denise. *New and Selected Essays*. New York: New Directions, 1992.
———. *Poems, 1972–1982*. New York: New Directions, 2001.
Levinas, Emmanuel. *Difficult Freedom*. Translated by Seán Hand. London: Athlone Press, 1990.
———. "Philosophy and the Idea of Infinity." In *Collected Philosophical Papers*, translated by Alphonso Lingis, 47–59. Dordrecht: Martinus Nijhoff, 1987.
———. *Totality and Infinity: An Essay on Exteriority*. Translated by Alphonso Lingis. Pittsburgh, PA: Duquesne University Press, 1969.
Libeskind, Daniel. *Radix-Matrix: Daniel Libeskind Architekturen und Schriften*. Edited by Alois Martin Mueller. Munich: Prestel, 1994.
Liska, Vivian. "'Roots against Heaven': An Aporetic Inversion in Paul Celan." *New German Critique* 91 (Winter 2004): 41–56.
Loraux, Nicole. *Né de la terre: Mythe et politique à Athènes*. Paris: Seuil, 2009.
Macauley, David. *Elemental Philosophy: Earth, Air, Fire, and Water as Environmental Ideas*. Albany: State University of New York Press, 2010.
Macfarlane, Robert. "The Word-Hoard: Robert Macfarlane on Rewilding Our Language of Landscape." *London Guardian*, February 27, 2015, www.theguardian.com/books/2015/feb/27/robert-macfarlane-word-hoard-rewilding-landscape.
Magris, Claudio. *Danube*. Translated by Patrich Creagh. New York: Farrar, Straus & Giroux, 1989.
Mairesse, Anne N. "Return to Monsieur Teste?, or 'What Is a Man Capable of?': Valéry, Anthropologist of Modernity." *Modern Language Notes* 117, no. 5 (December 2002): 1003–27.

Malkiel, Yakov. *Essays on Linguistic Themes*. Berkeley: University of California Press, 1968.

———. *Etymology*. Cambridge: Cambridge University Press, 1993.

Malpas, Jeff. *Heidegger's Topology: Being, Place, World*. Cambridge, MA: MIT Press, 2012.

Maly, Kenneth. "Reticence and Resonance in the Work of Translating." In *From Phenomenology to Thought, Errancy, and Desire*, edited by Babette E. Babich, 147–56. Dordrecht: Kluwer, 1995.

Marder, Michael. *The Philosopher's Plant: An Intellectual Herbarium*. New York: Columbia University Press, 2014.

———. *Plant-Thinking: A Philosophy of Vegetal Life*. New York: Columbia University Press, 2013.

———. "What Is Plant-Thinking? Botany's Copernican Revolution." *Los Angeles Review of Books*, March 17, 2013, http://lareviewofbooks.org/essay/what-is-plant-thinking-botanys-copernican-revolution.

Marx, Karl. *Critique of Hegel's Philosophy of Right*. Translated by Samuel Moore and Edward Aveling. Chicago: Aristeus Books, 2012.

———. "The Power of Money in Bourgeois Society." In *Economic and Philosophic Manuscripts of 1844*, translated by Martin Milligan, 165–69. New York: International, 1964.

Maupassant, Guy de. *The Horla*. Translated by Charlotte Mandell. Hoboken, NJ: Melville House, 2005.

Maurras, Charles. "La querelle du peuplier." *Paris Gazette de France*, September 14, 1903, 1–3.

McKenna, Andrew J. "Rorty, Girard, and the Novel." *Renascence* 55, no. 4 (Summer 2003): 292–313.

McLellan, David. *Utopian Pessimist: The Life and Thought of Simone Weil*. New York: Poseidon Press, 1990.

Mehlman, Jeffrey. "Writing and Deference: The Politics of Literary Adulation." *Representations* 15 (Summer 1986): 1–14.

Meringer, Rudolf. "Vorwort." *Wörter und Sachen* 1 (1909): 1–2.

Merleau-Ponty, Maurice. "Un inédit de Merleau-Ponty." *Revue de Métaphysique et de Morale* 67, no. 4 (October–December 1962): 401–9.

———. *Signes*. Paris: Gallimard, 1960.

———. *The Visible and the Invisible*. Translated by Alphonso Lingis. Evanston, IL: Northwestern University Press, 1968.

Miller, Christopher L. "The Postidentarian Predicament in the Footnotes of *A Thousand Plateaus*: Nomadology, Anthropology, and Authority." *Diacritics* 23, no. 3 (Fall 1993): 6–35.

Miller, Elaine P. *The Vegetative Soul: From Philosophy of Nature to Subjectivity in the Feminine*. Albany: State University of New York Press, 2002.

Miller, J. Hillis. "Derrida's *Destinerrance*." *Modern Language Notes* 121, no. 4 (September 2006): 893–910.

Miller, Paul Allen. "The Classical Roots of Poststructuralism: Lacan, Derrida, and Foucault." *International Journal of the Classical Tradition* 5, no. 2 (Fall 1998): 204–25.

Morgan, Robert. *Terroir*. New York: Penguin, 2011.

Morgenstern, Christian. *The Gallows Songs*. Translated by Max Knight. Berkeley: University of California Press, 1963.

———. *Werke und Briefe. Vol. 2, Lyrik 1906–1914.* Stuttgart: Urachhaus, 1992.

Mourelatos, Alexander P. D. *The Route of Parmenides: A Study of Word, Image, and Argument in the Fragments.* New Haven, CT: Yale University Press, 1970.

Mugerauer, Robert. *Heidegger and Homecoming: The Leitmotif in the Later Writings.* Toronto: University of Toronto Press, 2008.

Munro, Martin. "L'exil et l'innocence dans le discours politique et littéraire d'Haïti." In *Problématiques identitaires et discours de l'exil dans les littératures francophones*, edited by Anissa Talahite-Moodley, 131–50. Ottawa: Presses de l'Université d'Ottawa, 2007.

Nancy, Jean-Luc. *Dis-Enclosure: The Deconstruction of Christianity.* Translated by Bettina Bergo, Gabriel Malenfant, and Michael B. Smith. New York: Fordham University Press, 2008.

Nielsen, Dorothy M. "Prosopopoeia and the Ethics of Ecological Advocacy in the Poetry of Denise Levertov and Gary Snyder." *Contemporary Literature* 34, no. 4 (Winter 1993): 691–713.

O'Brien, Conor Cruise. "Patriotism and *The Need for Roots*: The Antipolitics of Simone Weil." In *Simone Weil: Interpretations of a Life*, edited by George Abbott White, 95–110. Amherst: University of Massachusetts Press, 1981.

Ovid. *Metamorphoses.* Vol. 1. Translated by John Dryden et al. London: Valpy, 1833.

Pacotte, Julien. *Le réseau arborescent, schème primordial de la pensée.* Paris: Hermann, 1936.

Palmer, Abram Smythe. *Folk-Etymology: A Dictionary of Verbal Corruptions or Words Perverted in Form or Meaning, by False Derivation or Mistaken Analogy.* London: George Bell, 1882.

Parker, Thomas. "Saint-Evremond and the Case of Champagne d'Ay: Early Modern French Aesthetic Theory Viewed through the Optic of *Terroir*." *Papers on French Seventeenth-Century Literature* 37, no. 72 (2010): 129–46.

Pascal, Blaise. *The Thoughts, Letters, and Opuscules of Blaise Pascal.* Translated by O. W. Wight. New York: Hurd & Houghton, 1869.

Paulhan, Jean. *La preuve par l'étymologie.* Paris: Editions de Minuit, 1953.

Pepper, Stephen C. "Philosophy and Metaphor." *Journal of Philosophy* 25, no. 5 (March 1928): 130–32.

———. *World Hypotheses: A Study in Evidence.* Berkeley: University of California Press, 1961.

Philo. *Works.* Vol. 1. Translated by C. D. Yonge. London: Henry G. Bohn, 1854.

Pierre, Noël. *Soleil Noir.* Paris: Seghers, 1950.

Pierrot, Jean. "La revanche du végétal de Zola à Caillois." In *Le végétale*, edited by Jean-Pierre Cléro and Alain Niderst, 251–86. Rouen: Publications de l'Université de Rouen, 1999.

Pinker, Steven. "Strangled by Roots: The Genealogy Craze in America." *New Republic,* August 6, 2007, 32–35.

Plato. *Timaeus and Critias.* Translated by Desmond Lee. London: Penguin, 2008.

Pollan, Michael. *The Botany of Desire: A Plant's-Eye View of the World.* New York: Random House, 2002.

Ponge, Francis. *Méthodes.* Paris: Gallimard, 1961.

———. *Nouveau nouveau recueil, 1940–1975.* Paris: Gallimard, 1992.

———. *Nouveau recueil.* Paris: Gallimard, 1967.

———. *Pour un Malherbe.* Paris: Gallimard, 1965.

———. *Tome premier*. Paris: Gallimard, 1965.

Prévert, Jacques. *Arbres*. Paris: Gallimard, 1976.

Quinn, Malcolm. *The Swastika: Constructing the Symbol*. London: Routledge, 1994.

Rasula, Jed. *This Compost: Ecological Imperatives in American Poetry*. Athens: University of Georgia Press, 2002.

Rorty, Richard. "Philosophy as a Kind of Writing: An Essay on Derrida." *New Literary History* 10, no. 1 (Autumn 1978): 141–60.

Rouart, Eugène. "Un prétexte." *L'Hermitage* 14, no. 12 (December 1903): 249–62.

Rousseau, Jean-Jacques. *Essay on the Origin of Languages and Writings Related to Music*. Hanover, NH: Dartmouth University Press, 1998.

Said, Edward W. *The World, the Text, and the Critic*. Cambridge, MA: Harvard University Press, 1983.

Saint-Exupéry, Antoine de. *The Little Prince*. Translated by Irene Testot-Ferry. Hertfordshire, UK: Wordsworth, 1995.

Sartre, Jean-Paul. *Being and Nothingness: An Essay on Phenomenological Ontology*. Translated by Hazel E. Barnes. New York: Philosophical Library, 1956.

———. *L'homme et les choses*. Paris: Seghers, 1947.

———. *Nausea*. Translated by Lloyd Alexander. Norfolk, CT: New Directions, 1949.

———. *Truth and Existence*. Translated by Adrian van den Hoven. Chicago: University of Chicago Press, 1992.

———. *War Diaries: Notebooks from a Phoney War, November 1939–March 1940*. Translated by Quintin Hoare. London: Verso, 1984.

———. *The Words*. Translated by Bernard Frechtman. New York: George Braziller, 1964.

Saussure, Ferdinand de. *Course in General Linguistics*. Translated by Roy Harris. LaSalle, IL: Open Court, 1986.

Scheler, Max. *Formalism in Ethics and Non-Formal Ethics of Values*. Translated by Manfred S. Frings and Roger L. Funk. Evanston, IL: Northwestern University Press, 1973.

Schleicher, August. *Darwinism Tested by the Science of Language*. Translated by Alex V. W. Bikkers. London: John Camden Hotten, 1869.

Schleiermacher, Friedrich. *Ästhetik*. Leipzig: Walter de Gruyter, 1931.

Schliemann, Heinrich. *Troy and Its Remains*. London: John Murray, 1875.

Schmidt, Albert-Marie. *La mandragore*. Paris: Flammarion, 1958.

Schor, Naomi. "The Crisis of French Universalism." *Yale French Studies* 100 (2001): 43–64.

Serres, Michel. *Hermes: Literature, Science, Philosophy*. Edited by Josué Harari and David Bell. Baltimore: Johns Hopkins University Press, 1982.

Sharr, Adam. *Heidegger's Hut*. Cambridge, MA: MIT Press, 2006.

Starobinski, Jean. "The Inside and the Outside." *Hudson Review* 28, no. 3 (Fall 1975): 333–51.

Steinbock, Anthony J. *Phenomenology and Mysticism: The Verticality of Religious Experience*. Bloomington: Indiana University Press, 2007.

Steiner, George. *Martin Heidegger*. New York: Viking Press, 1979.

———. *My Unwritten Books*. New York: New Directions, 2008.

Steiner, Rudolf. *The Foundation Stone*. Translated by Daisy Aldan. Forest Hills, NY: Folder Editions, 1987.

———. *A Modern Art of Education*. Great Barrington, MA: Anthroposophic Press, 1961.

———. *The Roots of Education*. Great Barrington, MA: Anthroposophic Press, 1997.

Straus, Erwin W. *Phenomenological Psychology*. Translated by Erling Eng. New York: Basic Books, 1966.

Struever, Nancy. "Fables of Power." *Representations* 4 (Autumn 1983): 108–27.

Syrotinski, Michael. "Domesticated Reading: Paulhan, Derrida, and the Logic of Ancestry." In *The French Connections of Jacques Derrida*, edited by Julian Wolfreys, John Brannigan, and Ruth Robbins, 85–100. Albany: State University of New York Press, 1999.

Taine, Hippolyte. *Introduction à l'histoire de la littérature anglaise*. Boston: Heath, 1898.

Tally, Robert T., Jr. "Nomadography: The 'Early' Deleuze and the History of Philosophy." *Journal of Philosophy* 5, no. 11 (Winter 2010): 15–24.

Taylor, John. *Paths to Contemporary French Literature*. Vol. 2. New Brunswick, NJ: Transaction, 2009.

Taylor, Paul Beekman. "Some Uses of Etymology in the Reading of Medieval Germanic Texts." In *Hermeneutics and Medieval Culture*, edited by Patrick J. Gallacher and Helen Damico, 109–20. Albany: State University of New York Press, 1989.

Tobias, Rochelle. *The Discourse of Nature in the Poetry of Paul Celan: The Unnatural World*. Baltimore: Johns Hopkins University Press, 2006.

Tournier, Michel. *Friday*. Translated by Norman Denny. New York: Pantheon, 1969.

———. *The Mirror of Ideas*. Translated by Jonathan F. Krell. Lincoln: University of Nebraska Press, 1998.

———. *The Ogre*. Translated by Barbara Bray. Baltimore: Johns Hopkins University Press, 1997.

———. *Vendredi, ou les limbes du Pacifique*. Paris: Gallimard, 1972.

Tribout-Joseph, Sarah. *Proust and Joyce in Dialogue*. London: Legenda, 2008.

Tull, Jethro. *The Horse-Hoeing Husbandry; or, A Treatise on the Principles of Tillage and Vegetation, Wherein Is Taught a Method of Introducing a Sort of Vineyard Culture into the Corn-Fields, in Order to Increase Their Product and Diminish the Common Expense*. London: William Cobbett, 1829.

Valéry, Paul. *Cahiers*. Vols. 1–29. Paris: CNRS, 1957–62.

———. *Dialogues*. Translated by William McCausland Stewart. London: Routledge, 1957.

———. *Monsieur Teste*. Translated by Jackson Mathews. Princeton, NJ: Princeton University Press, 1973.

———. *Œuvres*. Vol. 2. Paris: Gallimard, 1960.

Ventresque, Renée. "Des bienfaits de l'étymologie: Deux lectures de Saint-John Perse à l'époque de la création d'*Amers: La preuve par l'étymologie* de Jean Paulhan et *Mallarmé* de Wallace Fowlie." *Souffle de Perse* 2 (January 1992): 56–64.

Voegelin, Eric. "In Search of the Ground." In *The Collected Works of Eric Voegelin, vol. 11, Published Essays, 1953–1965*, 224–51. Columbia: University of Missouri Press, 2000.

Wandruszka, Mario. "Etymologie und Philosophie." In *Etymologica: Walther von Wartburg zum siebzigsten Geburtstag*, 857–71. Tübingen: Max Niemeyer Verlag, 1958.

Weil, Patrick. *Qu'est-ce qu'un Français? Histoire de la nationalité française depuis la Révolution*. Paris: Grasset, 2002.

Weil, Simone. *Ecrits historiques et politiques*. Paris: Gallimard, 1960.

———. *Gravity and Grace*. Translated by Arthur Wills. Lincoln: University of Nebraska Press, 1997.

———. *The Need for Roots*. London: Routledge, 2007.

———. *Notebooks*. 2 vols. Translated by Arthur Wills. New York: G. P. Putnam's, 1956.

———. *Waiting for God*. Translated by Emma Craufurd. New York: G. P. Putnam's, 1951.

Weltman-Aron, Brigitte. "Rhizome and Khôra: Designing Gardens with Deleuze and Derrida." *Bulletin de la Société Américaine de Philosophie de Langue Française* 15, no. 2 (Fall 2005): 48–66.

Wortham, Simon Morgan. *The Derrida Dictionary*. London: Continuum, 2010.

Worton, Michael. "*Waiting for Godot* and *Endgame*: Theater as Text." In *Samuel Beckett's "Waiting for Godot,"* edited by Harold Bloom, 71–92. New York: Infobase, 2008.

Ziolkowski, Theodore. "Twentieth-Century Variations on *Eclogues* 1." In *Pastoral Palimpsests: Essays in the Reception of Theocritus and Virgil*, edited by Michaíl Paschális, 155–70. Iráklion, Crete: Crete University Press, 2007.

INDEX